Safety Management: Strategy and Practice

Safety Management: Strategy and Practice

Roger Pybus

Butterworth-Heinemann Ltd
Linacre House, Jordan Hill, Oxford OX2 8DP

℞ A member of the Reed Elsevier plc group

OXFORD   LONDON   BOSTON
MUNICH   NEW DELHI   SINGAPORE   SYDNEY
TOKYO   TORONTO   WELLINGTON

First published 1996

© Butterworth-Heinemann 1996

British Library Cataloguing in Publication Data
A catalogue record for this book is available from the British Library
ISBN 0 7506 2519 8

Library of Congress Cataloguing in Publication Data
A catalogue record for this book is available from the Library of Congress

Composition by Genesis Typesetting, Rochester, Kent
Printed in Great Britain by Clays, St. Ives PLC

Contents

Preface

Welcome to *Safety management: strategy and practice*, which I offer not as a thesis on the detailed theory or social virtues of safety – there are other learned and highly technical texts which do this – but as a practical route map of good practice in safety management for sound business purposes. Although it is *'safety'* that appears in the title, and throughout the text, the application is actually broader, embracing the *health* of people at work – and of the communities in which industry and commerce operate – and the management of the *environmental* impact of business operations. Getting any of these key operating features badly wrong has seriously disabled many businesses – some of them major international players. So, where you read 'safety', think safety, health and environment, and indulge my desire to save my pen, and to avoid lengthening the text any more than is necessary.

Of course, the social and moral aspects of safety cannot be ignored. They are intimately tied into corporate values about people and work, recognizing that the impact of poor safety management on the morale and motivation of people at work ultimately has a similar bottom line impact on overall company performance and profitability to the direct and consequential losses from accidents themselves. Of course, the driving forces are not just internal. Modern safety, health and environmental legislation is increasingly demanding, reflecting the growing demands of the public and high profile pressure groups, for industry to minimize any adverse impact from its activities and products on the consuming public and the environment.

For those of us in industry, it is both a major challenge and an opportunity to gain competitive advantage from being strategically and technologically smarter and better than the competition in the way we meet these growing safety, health and environmental demands.

Learn to manage safety well and you will have learned how to manage *anything* well. Good safety management is the same as good *anything* management. If you are in any doubt, just ask the best performers in the business.

Roger Pybus

Acknowledgements

The text contains much that I have learned from safety, health and environment professionals and enthusiastic managers, both in my own industry and elsewhere. They are too many to mention, but they will know who they are – I acknowledge you all.

Special thanks to Bob for his professional shaping of the chapter on training strategy, and to Sue and Maureen for their patient advice on illustrations.

Chapter 1

The strategic role of safety management in modern business practice

'Of all God's creations, corporations seem to have the shortest memories of all'

Let's face it, being in business is not simple. There are all the complexities of forecasting market trends, keeping and attracting new customers, maintaining productive assets, co-ordinating sales and deliveries . . . and much more. So, isn't it inevitable, given all this complexity, that there will be things that go wrong? Things that result in accidents? Things that result in a little damage to people and other assets? Sometimes, things that result in serious injury or damage? Isn't that inevitable?

Let's face it, safety costs money, doesn't it? It brings extra bureacracy and slows things down, doesn't it? All that red tape; all those constraints on getting on with the *real* job.

Safety. Not a good fit with the modern business image of risk taking, cost cutting and enterpreneurial spirit, is it?

Familiar arguments? You've probably heard them before. You may have used them yourself. But how do those arguments stand up against the facts? Let's explore. First, we'll take a quick look at accidents – what's to predict, and what's to prevent. Then we'll ask ourselves who cares.

In any organization, unless safety management is really excellent, important things will be going wrong. These failures usually surface as relatively minor events – a slight injury from a fall, damage to a machine, a minor transport accident, a small electric shock, a small fire. Sometimes, they don't even result in injury or other damage, but are simply a 'near miss' – a swerve in a car, a trip on a loose cable, a roof tile falling onto a walkway, a motor found sparking in a flameproof area.

These are the sort of incidents that don't get much attention. They often don't interrupt the work, and so are dismissed and forgotten – maybe never even reported. All part of the ups and downs of life, usually put down to personal carelessness.

Then there are the slightly more serious incidents. Those that result in enough damage to significantly interrupt normal operations: a fire that damages processing equipment, putting it out of action for a time; a fractured wrist from a fall off a platform which reduces your operating team for a week or two; a punctured drum of toxic material which leaks into a nearby watercourse killing fish in a local angling spot; a car accident involving one of your sales team, resulting in a few late or lost orders during his time off work. These incidents get more attention because they upset the normal workflow.

Are the underlying causes of these incidents very different from those resulting in minor accidents? Not at all.

Then there are the even more serious incidents: a fire leading to key production plant shutdown for a few days; children suffering chemical burns from a leak of corrosive chemical which escapes outside the premises; delivery of the wrong substance to a customer, resulting in spoilage of their product; serious injury to an employee leading to a restraining order by the Health and Safety Executive. Now we are talking about incidents which can result in business interruption and constraints that are really starting to affect profitability and confidence amongst the communities in which we are operating. Are the underlying causes of these incidents significantly different from those resulting in minor accidents? Not at all.

Then there is the 'chance in a million' when something disastrous happens: explosion, fire, major release of harmful substance into the local community or environment, key executives killed in a travel accident. The sort of accident from which a business may never fully recover – which, in the worst case, will put you out of business altogether. Are the underlying causes of these incidents significantly different from those resulting in minor incidents? You guessed it – not at all.

What level of incident, then, are we prepared to accept, and which are we not? Can we accept the minor incidents without accepting the major incidents? Well, you could argue that the big incidents are very much less frequent than the smaller ones. That's true; the odds of having a major incident are longer, but like any lottery, how do you know that million to one chance will occur tomorrow, next week, next year or never? You don't. That big event could be 50 years or 50 days away.

A little later we'll look at some incidents and what they tell us, both about their underlying causes and about their predictability. Why, if the underlying causes of incidents – big and small – are predictable, are their consequences so unpredictable? Actually, it's quite straightforward. Every incident resulting in loss (injury or damage) is the product of more than one contributory – and often variable – factor.

Take a simple example. A driver, driving too fast, brakes hard on a corner and his car skids across the road. What are the potential consequences? One consequence is that there is no loss – there is no oncoming traffic and the driver manages to correct the car back to his side of the road without any injury to himself or damage to the car. A temporary increase in heart rate, no doubt, but that's all. An alternative consequence is that there is an oncoming lorry and the driver is killed. You could think of many intermediate consequences between these two extremes. The consequences are unpredictable because they depend on the circumstances at the time – the 'contributory factors'. In this case, these are:

- the speed of the car
- the condition of the road surface
- the tightness of the bend
- the response of the driver
- the presence or absence of roadside obstacles and oncoming traffic.

You might identify more. Think about the nature and variability of these contributory factors. Some are conditions (the second, third and fifth), and some are acts. All but the third are variable. The point is, there are many combinations of these circumstances, only some of which will result in serious consequences (see Figure 1.1).

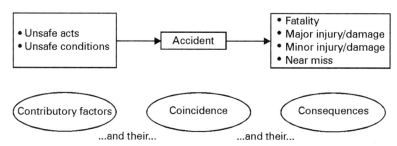

Figure 1.1

If we decide that serious consequences are unacceptable, how do we avoid them? We can do this in one of two ways – by accepting the contributory factors (and the possible consequent incident) and taking action to minimize the consequences, or by removing enough of the contributory factors to avoid the accident happening in the first place.

The first may not always be practicable, and can often be expensive. Given the worst case in the example above, to avoid serious injury we would have to design a damage-proof car, which would be an unreasonably expensive solution – unless you know a good supplier of second-hand armoured vehicles! Let's explore the second option – to influence the contributory factors. Look back at them. Which of them are controllable and which not?

There are two controllable factors – the speed of the car and the response of the driver – both of them relating to driver behaviour. Improving driver behaviour (by training in defensive driving and good roadcraft, for example) will significantly reduce the risk of accident, and at very little expense.

Let's look at the example above in a slightly different way. If we mapped out as a time sequence the combination of factors that could lead to a serious accident, it might look something like Figure 1.2. You will see that some of these factors coincide from time to time, and when they do there is sometimes a near-miss accident (wet road, driving too fast, poor response resulting in braking too late on a bend, and loss of control). When they *all* combine, however (all the factors above *and* oncoming traffic), a serious accident may occur.

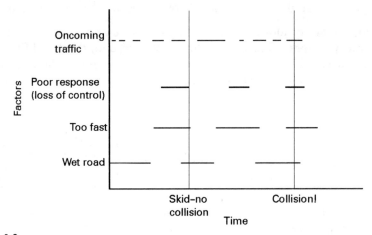

Figure 1.2

Predicting how often this is likely to occur can be tricky, but can – given data based on experience or actual incidents – be estimated. We explore this whole area of risk assessment and management in more detail in Chapter 4.

When there is significant loss of life as the result of an accident in the UK, there is usually a judicial enquiry, which examines the circumstances of the accident in great detail. The lessons are invariably very instructive, so we will take a look at a few in some detail, and see what they tell us about the things we've just covered – underlying causes (the contributory factors), coincidence and consequences.

Clapham Junction

On 12 December 1988, a crowded commuter train ran head-on into the rear of a stationary train outside Clapham Junction station. The impact of the collision caused the first train to veer off and strike an oncoming train. Thirty-five people died and nearly 500 more were injured, 69 of them seriously.

The incident happened between the last two signals before the station. The signalling system had failed because during alterations to the signalling system a wire should have been removed, but it had not. It made electrical contact with its old circuit, enabling a flow of current into the new circuit, when that circuit should have been dead. The flow of current prevented the signal from turning to red – the 'stop' signal.

Preparation of the new wiring was done during the week and connections made at the weekend. Part of the rewiring required an old wire to be removed from its two end terminals, the fuse end and the relay end. Both ends should have been disconnected from their terminals, cut back to ensure no possible future physical contact, the wire ends taped over and then tied back. The fuse end was left connected. The relay end was disconnected but not cut back and not taped over. During another weekend job adjacent to this relay, *two weeks* later, the wire was disturbed and came back into contact with the relay.

What were the lessons?

The *immediate* cause (the failure to cut back and tape the old wires so that they could not interfere with the new circuitry) was very simple and obvious. So how, you might ask, could an experienced technician fail to follow such an important safe practice, when the potential consequences – faults in the train signalling system – were so serious? And how, bearing in mind the seriousness of the consequences, did this error go unchecked?

The underlying causes tell us clearly.

Cause 1 There was lack of clarity about job responsibilities and accountabilities

The senior technician in direct charge of carrying out the rewiring had no job description, nor any training as a supervisor, a role he was sometimes expected to perform. *His* supervisor was so involved on the tracks leading a gang of men in outside work, and often taking part in that work himself, that he totally neglected his role as a supervisor. Those further up the management chain who were responsible for ensuring that an independent wire count was carried out on safety relays and terminations were not aware of their responsibilities, thinking that such checks were the responsibility of those carrying out and supervising the actual work.

Cause 2 Organizational restructuring failed to take account of the resources needed to make the new structures work effectively

Senior management set requirements on the Signal Works Engineer for testing of rewiring work prior to commissioning without any thought to detail or resources. The Signal Works Engineer left the test and commissioning engineer to get on with the testing without any checks or feedback. Right down the management chain there was an assumption that everything would be done as required, with little or no understanding of the resources needed to meet the requirements for drawing up, carrying out and testing the specified rewiring work.

Cause 3 There was a failure of management leadership and communication

Major reorganization had taken place, but was not accompanied by reasonable consultation, planning, communication and training, and by adequate new job descriptions. Communication down the management chain was very haphazard; below a certain level in the management chain, formal communication ceased. The instruction which described the requirement that 'a wire count must be carried out on all free-wired safety relays and terminations and recorded on the contact/terminal analysis sheets' did not reach the supervisors who were responsible for carrying out the instruction at the working level. This important work instruction was issued with no accompanying explanation by management, and there was no training in how it should be implemented, despite its having taken three years to prepare and issue. Documentation which should have accompanied the instruction, including checklists, was never issued.

Cause 4 Training was inadequate

The senior technician who was responsible for the rewiring which resulted in the accident was regarded, in the eyes of his colleagues and his line managers, as competent and efficient. Yet many of the errors that he made which resulted in the signalling failure that caused the accident, he had been making all his working life. Rather than being an isolated lapse, they had become his standard working practices. Nor were these malpractices confined to him – they were part of a widespread way of working at technician, senior technician and supervisor level. He himself had had minimal training, and had no technical qualification at all for the job he was doing. He had received no training in the requirements of the instruction to cut back, tape and tie back disconnected wires so that they could not remake contact with the terminals from which they had been disconnected. He had no concept of how essential it was to carry out these actions, or of the need for an independent wire check by a third party not involved with the work. And, largely because of limited funding for formal training, 'experienced' technicians such as the senior technician in question were not expected to take formal training, and there was no refresher training for those doing his type of work.

Cause 5 Established safe systems of work were not followed

There were many examples of established systems of work having been ignored. Here are some examples:

- the proper procedure for specifying rewiring details was frequently not followed
- 'on the job' modifications were made to wiring details where practical problems in following the design details arose and, although these were subject to informal consultation with the design office, the formal rewiring drawings were not amended
- no independent wire counts were carried out, as required
- redundant wires were not disconnected and tied back, as required.

Cause 6 There were failures to audit

Safety auditing was not part of the organizational culture. As a result, not only were there no rewiring checks, as required by the instruction about independent wire counts, but there was no auditing system to check whether the wire counts were taking place. No checks and balances and no opportunity for identifying that things were not being done as they needed to be.

Cause 7 The warning signals from previous incidents were ignored

Three years prior to the incident, there had been very similar signalling failures following rewiring work on another part of the regional railway system. After the commissioning work was done, there were three incidents. In one of them two wires had been wrongly connected to the same terminal of a relay when there should have been only one wire. The parallels to the Clapham Junction incident are striking; the lessons were not learned.

Reflecting on these underlying issues, it is not so surprising that the incident occurred, when people did not know clearly their job responsibilities and accountabilities and did not receive adequate training, when poor leadership and communication resulted in a disconnected organization with no effective checks on management systems, and when the warnings from past incidents were studiously ignored.

Piper Alpha

On 6 July 1988, on the oil rig Piper Alpha, 100 miles off the east coast of Scotland in the North Sea, there was an escape of flammable gas. The gas ignited, causing an explosion and fire and 167 people died.

The estimated cost of the incident to the Occidental Group which owned the platform, including consequential insurance payouts, is over £2 billion.

The *immediate* cause of the gas escape was the putting into service of a condensate (liquefied flammable gas) injection pump, downstream from which a pressure relief valve had earlier been removed for maintenance. Liquefied flammable gas escaped under pressure from the site from which the valve had been removed. Those who brought the pump back into service were not aware of the work on the line downstream. How, you might ask, were they not aware of this other work which had such obvious, direct and significant effect on their own actions? Where were the control systems to prevent this happening?

The underlying causes tell us clearly.

Cause 1 There was no adequate risk assessment

The consortium which owned the oil rig, the Occidental Group, had, as required under section 2(3) of the Health and Safety at Work, etc. Act, 1974, a statement of safety and health policy which included an assertion that 'The promotion of health and safety is an integral part of the duties of line management and should be afforded the same priority as other key responsibilities'. Bearing in mind the hazardous nature of its operations, rigorous hazard studying of the installation might have been expected, for although there was no statutory requirement for British offshore installations to have a formal assessment of operational safety, it was the practice of others in the industry to carry out such assessments, similar to those required for onshore major hazard installations. In fact, the extreme consequences of a prolonged high pressure gas fire *were* recognized, and the cause of such a fire – failure and depressurization of a gas pipeline – identified. However, a comprehensive analysis of hazards was not carried out and, whilst there was provision for fire separation to contain the outbreak of fire, there was no provision for explosion. When the explosion occurred, therefore, it destroyed the fire separation, allowing the fire to spread.

Another factor, not so much associated with the initial explosion as with the ensuing consequences, was that the two other platforms which interconnected with Piper Alpha were not prepared for an emergency on another platform, and failed to stop their oil production when they were made aware of the explosion and fire on Piper Alpha. Had they taken this action, the second major explosion on Piper Alpha – due to the rupture of the riser on one of the gas pipelines as a result of the intensity of the oil fire being fed by production on the other platforms – would probably have been delayed.

Cause 2 The permit to work system was inadequate

A permit to work system is a formal written system for ensuring that the hazards associated with potentially hazardous jobs are adequately considered, and formal controls of those hazards identified so that a safe system of work for the job may be specified. The permit to work system operating on Piper Alpha failed to ensure a consistent system for warning tags on isolation valves which had been closed as part of the isolation of equipment for maintenance. If such a system had been operative, the downstream isolation valve on the condensate pump which was put back into service, resulting in the flammable gas escape, would have been tagged. There were also inadequacies in the arrangements for cross-referencing permits to work where the work controlled by one permit could affect the work controlled by the other, and in the arrangements for handover of permits at shift changeover.

Cause 3 The permit to work system was not followed

The enquiry exposed that the permit to work system was commonly not followed. Typical failures ranged from failure to insert dates, times and signatories, failure to electrically isolate, failure to describe the nature of the work adequately, failure to complete sections of the permit to work form, failure to display the permit at the work site, failure to hand over permits personally . . . and more. Operating staff had no commitment to the written procedure.

The following two causes indicate the underlying reasons – the actions of management did not give any indication that following the system was important.

Cause 4 There was no formal training in the permit to work system

Occidental provided no formal training in the permit to work system, so that those required to operate it had to pick it up from watching others who were authorized to issue permits. Not surprisingly, custom and practice overtook the procedure itself. This *ad hoc* arrangement applied equally to contractors, who were expected by Occidental to ensure that their employees were familiar with the system.

Cause 5 Auditing of the permit to work system was inadequate

The permit to work system was a critical part of the arrangements to ensure safe operation on the platform. Yet the arrangements for monitoring the system were inadequate. There was no formal procedure as to how safety auditing should be carried out, and management involvement in the auditing process was limited. The auditing that *was* carried out was superficial, and failed to discover the deficiencies in the operation of the permit to work system. Even a corporate audit carried out in the last quarter of 1987, which looked at permits sent from Piper Alpha, had failed to report any deficiencies.

Cause 6 Senior managers assumed that 'no news is good news'

It was apparent from the evidence at the enquiry that line managers assumed that other managers lower down the management chain were carrying out audits of the permit to work system satisfactorily, and that the lack of any feedback on any permit to work problems indicated that all was well.

The reality of no news is that the bad news is not getting through.

Cause 7 There was failure to act on known deficiencies

An important feature of operational safety on an oil platform is firefighting capability. The reason for this lies in the difference between on-shore and off-shore installations; on-shore the operators can escape from a fire by running away, but on an oil rig this is not an option. Reliable firefighting arrangements are crucial, and it was Occidental's approach to fire protection that fire be controlled before heat damage could occur to oil and gas bearing equipment, or to the structural members of the platform. This was especially critical on Piper Alpha because it had no structural fireproofing and the structural members were under high stress. Even so, the fire pumps which supplied water to the firefighting systems were taken off automatic for significant periods of time – those times during which diving work was taking place under the platform. There was an imbalance in the assessment of risk between diver safety and loss of automatic firefighting water.

The other shortcoming in the firefighting arrangements was that many of the deluge system spray heads were blocked by scale resulting from internal corrosion of the galvanized carbon steel deluge distribution pipework by sea water. This problem was identified as early as 1984. The only practicable solution – to replace the carbon steel pipework with stainless steel – was significantly delayed, even though it had a high priority.

Cause 8 The warning signals from earlier incidents were ignored

In September 1987 a contract rigger was killed in an accident on Piper Alpha, primarily as a result of inadequate specification of a safe system of work that was being controlled by a permit to work. These inadequacies did not result in a serious questioning of the overall effectiveness of the arrangements to ensure safe systems of work, of which the permit to work is the central part.

All these causes tell us quite clearly where the control systems were – our opening question on this incident. They were inadequate, they were not followed, they were not audited effectively, no-one took any action on things that were known to be wrong, and when these same deficiencies contributed to the death of an operator, no effective action was taken to deal with them.

Space Shuttle *Challenger*

On 28 January 1986, at 11.38 Eastern Standard Time, the Space Shuttle *Challenger* was launched from the Kennedy Space Center. Seventy-three seconds later its flight ended in an explosive burn of hydrogen and oxygen propellants. All seven crew members died.

The immediate cause of the disaster was the failure of a rubber O-ring on one of the two solid fuel booster rockets, leading to a leak of fuel, which ignited. The flame deflected on to the surface of the external liquid hydrogen/liquid oxygen Shuttle propellant tank, causing failure of the tank structure and escape of hydrogen, adding to the fire. Structural failure followed, the escaping liquid fuels ignited explosively, and the Shuttle assembly was destroyed.

With all this space-age technology, how was it that a simple O-ring failed with such disastrous consequences? The Presidential Commission of enquiry unearthed significant underlying causes.

Cause 1 The design of the booster rocket section sealing arrangements was flawed

The dynamic characteristics of the booster rocket section O-ring seals were complex, and reliance on maintaining an adequate seal during lift-off was dependent on many variables. The design was intrinsically too sensitive to hardware variables, propulsion stresses and environmental influences such as air temperature and rainwater ingress. It was because of these constraints on the effectiveness of the O-rings in retaining an effective seal that the manufacturer's recommendation was that air temperature at launch should be no lower than 53 degrees Fahrenheit. The actual temperature at launch was 36 degrees Fahrenheit. There was a reliance on internally generated pressure in the booster rocket to energize the seal, and at the low launch temperature the reduced resilience of the rubber O-ring prevented its sealing effectively under the pressure generated in the booster rocket, allowing hot gases to flow past it and damage its integrity.

It is worthy of note that of the four competitors for the design and construction of the space shuttle's solid rocket booster, the evaluation board rated the successful competitor fourth on design, development and verification. A major factor in the success of this competitor appeared to have been cost.

Cause 2 There were shortcomings in hazard assessment, and warning signs were ignored

Despite the eventual awareness of the shortcomings of the design of the booster rocket section joints, there was no objective assessment of the risks associated with joint failure. NASA did not accept the judgement of its engineers that the design was unacceptable, and as the evidence of joint problems grew, they were minimized by NASA. As tests and subsequent flights confirmed that damage to the rings from hot gases was a significantly common occurrence, and therefore a significant risk to the safety of shuttle launch, the response both of NASA and the seal manufacturer was to broaden the criteria for acceptable damage. They had both become locked into defending a design that was not adequate as the project progressed and the cost of going back to redesign increased. All objectivity with regard to the design was lost, and the evidence and the risks ignored.

Cause 3 There were failures to communicate

As the problems with the O-ring became known, they were not communicated to the senior management team responsible for the decision to launch. This team was also unaware of the initial written recommendation of the O-ring manufacturer advising against launch at temperatures below 53 degrees Fahrenheit, and of the internal opposition of the manufacturers' engineers to their management's reversal of this recommendation. The problems were essentially hidden from those responsible for the launch.

The communication arrangements were weakened by a decision, taken by the then manager of the National Space Transportation System, to restrict the previous requirements on the team responsible for the programme of preparation for the launch for reporting upwards *all* problems, trends and problem close-out actions to only those problems which dealt with 'common hardware items or physical interface elements'. This revision eliminated the requirement to report on flight safety problems, flight schedule problems and problem trends, with the result that the senior pre-flight readiness review team were divorced from safety, operational and flight schedule problems.

Cause 4 There was pressure to achieve the launch

The launch of Space Shuttle *Challenger* was a very high profile, very high cost project, and the pressure to meet the original expectations of the project – to make space operations using the shuttle routine and economical – were great. As problems in the project built up, the pressure increased, and NASA's bullish approach to meeting difficult challenges added to the pressure. Resources were stretched and fundamental principles overridden. The determination to launch was so strong that there was a burden of proof that it was *not* safe to launch, rather than a burden of proof that it *was* safe to launch.

So, to answer our opening question as to how it was that a simple O-ring failed with such disastrous consequences – it was poor initial design that was not seriously challenged when its deficiencies became increasingly apparent, it was the emasculation of arrangements to report flight safety problems, and it was the replacement of a safety-driven culture by a performance-driven culture.

The *Herald of Free Enterprise*

On 6 March 1987 the roll-on, roll-off passenger and freight ferry *Herald of Free Enterprise* sailed from Zeebrugge for Dover in good weather with 459 passengers and cars and freight vehicles. Four minutes after having cleared the inner harbour, the ferry capsized with the loss of 188 lives and many other injuries.

The immediate cause of the accident was the ingress of seawater into the cargo deck, the inner and outer bow doors of which had not been shut. How could such a fundamental mistake have been made?

Cause 1 A general instruction to close the bow doors was regularly ignored

The ship's instructions, endorsed by the captain, included the requirement for the officer loading the main vehicle deck to ensure that the watertight bow and stern doors were secured when leaving port. However, the captain accepted that compliance was adequate if the responsible officer, before he left the main vehicle deck, ensured that the assistant bosun was at the control position (for closing the doors) and was *apparently* going to operate the doors. This was quite contrary to the instruction to ensure the bow and stern doors *were* closed. And because this non-compliance was accepted by the captain, it set the scene for non-compliance further down the management chain.

Cause 2 The ship's standing orders were not adequate for the load/unload arrangements at Zeebrugge

The *Herald of Free Enterprise* had two vehicle decks which, at Dover and Calais, could be loaded simultaneously. Zeebrugge harbour, however, was designed for loading on to the bulkhead deck of single deck ferries. For this (and other) reasons, it took longer to load at Zeebrugge, so that different arrangements for loading and turn-round were needed compared with those at Calais. No such arrangements had been made, and as a result there was pressure on the officer responsible for ensuring the vehicle loading doors were closed to get to his harbour station on the bridge as soon as possible.

A 'bridge and navigational procedures' guide issued by the company implied that the officer on watch should be on the bridge 15 minutes before sailing. When the officer on watch was the loading officer, the order created a conflict of interest – he could not be in two places at once.

Cause 3 There was no safe system of work

The general instruction to close the bow doors was the only safeguard to ensure the doors were closed before leaving harbour. We saw earlier that this was not enforced, and there was no back-up system to make sure the doors were closed before sailing, such as a simple 'door open' warning light on the bridge. In reality, there was no effective safe system of work to ensure the closure of the loading doors before sailing.

Cause 4 There was pressure from the company to turn the ship round, especially at Zeebrugge

There was considerable pressure to turn the ferries around quickly. This was especially true at Zeebrugge, exacerbated by the single-deck loading arrangements, and the demands to make up time from late departures at Dover because of the lack of time there to off-load and load. The company operations manager at Zeebrugge made it quite clear that he wanted the ship to sail 15 minutes early from Zeebrugge, and that pressure be applied to staff to achieve this target. This pressure, and the conflict of duties in the company general instruction described earlier, made the failure of the loading deck officer to ensure closure of the doors almost inevitable.

Cause 5 'Not my responsibility' attitude of staff

The assistant bosun, whose responsibility it was to close the bow doors, went to his cabin after having been released from his work by the bosun, and then fell asleep. The bosun, the last person to leave the loading deck before the boat sailed, did so without there being anyone there to close the doors. He took the view that it was not his job to make sure someone was there to close the doors, and that he was responding to the 'harbour stations' order by going to his station. The duties of the second officer, who was on the loading deck, were taken over by the Chief Officer, who failed to ensure the bow doors were closed. No-one took responsibility for checking that the doors were shut.

This failure to take responsibility had its roots further up the management chain, where there was a failure to comply with Merchant Shipping Notice requirements to provide the masters of ships with clear instructions for the safety and technical aspects of the ship's operation, and to provide clear role definitions for the senior members of the ship's management.

Cause 6 'No news is good news' philosophy

The company instructions for making ready to sail required the reporting of any deficiency likely to cause lack of readiness to sail to the ship's master, and that in the absence of such a report the master should assume that the ship was ready to sail. This made an assumption that if nothing was reported, all was well – a very dangerous premise, as we saw in the Piper Alpha case.

Cause 7 Failure to heed past incidents

The failure to close the bow doors on the *Herald of Free Enterprise* was not the first time such a failure had occurred. A few years earlier the assistant bosun of a sister ship had fallen asleep and not heard 'harbour stations' being called, with the result that both the bow and stern doors were not closed before sailing from the ship's berth at Dover. There had, in fact, been four other occasions when one of the company's ships had proceeded to sea without having shut the bow or stern doors. Some of these incidents were known to management, who had not brought them to the notice of other ships' masters. Nor, as a result of these known and significant factors, did the 'ships standing orders' – which were issued by the company – make any reference to opening and closing the bow and stern doors.

There were other failures of important marine practice, the sum total of which made the disaster eventually inevitable. It was just a matter of time.

Major incidents, simple failures. What are the lessons? Here are some obvious ones, in summary.

- Ignore the small accidents at your peril: they are the signals of worse to come.
- You have to know the worst things that can happen, and the risk of their happening.
- You have to have formal controls over those things you can't afford to go wrong.
- You can't expect people to do the right thing unless they know what to do, and have the resources to do it.

- You can't expect people to do the right thing if there are conflicts of interest. They have to know you are serious; you have to show it by your own behaviour.
- You will never know how well or how badly things are working unless you check carefully, and make it quite clear you want to hear the bad news as well as the good.
- Without clear lines of responsibility and accountability, things will be out of control.

Well, as I said at the beginning of this opening chapter, being in business is not simple. Staying in business depends a lot on how we handle the points above. How about your business? Do you, for example, know the worst thing that could happen in your business operations? Do you know what the risk is of that happening? Do you have clear lines of accountability for all the important safety (and other) aspects of your operations? Are you confident that you hear both the good news and the bad? Does your behaviour tell others you want to hear about the bad news, or does your explicit (or implicit) behaviour threaten people enough for them to cover it up? Well – how about it?

In the following chapters, we will explore the issues above in more detail, and we will refer back to the salutory, yet repeatedly learned lessons of these major tragedies.

Chapter 2
Making the change

'Plus ca change ...'

Because *systems* and *people* span the whole range of business activity, it is not unusual to see them being worked on in an *ad hoc* way – a little systems improvement here, a little people improvement there. The reasons are quite understandable: only so much work – only so much change – can be reasonably undertaken at any one time. However, the most effective way of exploiting this effort is within an overall strategic plan for improvement, the purpose and direction of which is understood by everyone. All efforts to improve safety will have an impact. Sometimes the impact is short-lived, particularly if it has little influence on people's thinking or behaviour, or if it is perceived as being disconnected from other, more mainstream activities. Other times, the impact is more permanent, particularly where it achieves a shift in understanding, attitudes and associated behaviours.

An interesting observation of companies with a world class safety performance is the seamless link between systems and people. The systems are a good fit with the organizational culture and the way people work, and people take the systems seriously and seek to improve them in a continuous cycle of dynamic growth. Many organizations fail to achieve this balance and synergy, focusing largely on improving systems. Of course, a focus on systems will produce results – but limited ones. For example, it is characteristic of organizations which are some way along the journey of improving safety that their performance improves, then reaches a plateau from which it is difficult to improve further (Figure 2.1). At this stage, something significantly different is needed to continue the gain in improvement.

This is a stage at which there can be significant management frustration – here we are, writing procedures, training people, even auditing, perhaps, and we are standing still. It's a bit like seeing flat sales figures when the sales team are doing their best to drum up new customers.

In this situation, it is very tempting to fall back onto campaigns, or similar one-shot initiatives. And, it has to be said, they usually achieve a quick

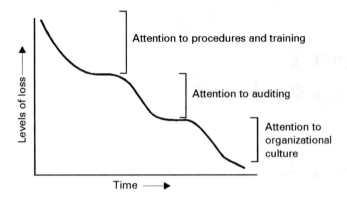

Figure 2.1

impact on performance. Injury levels drop, prizes are given, everyone feels gratified. But somehow, that improvement in performance always seems to be relatively short-lived. After the campaign finishes, the injury levels begin to rise again, back to their old levels – or maybe even worse. Is the answer to get back into another campaign? How many campaigns before people become saturated? And what does reliance on campaigns tell us about our underlying capability to manage our safety performance?

The worst possible example of this campaign approach is giving big prizes for good safety performance. What actually happens? Well, you can guarantee that performance will improve – but only temporarily, whilst you are offering prizes. During this time people are still having the same level of accidents, but they are hiding them so as not to compromise their chance of a prize. Even worse, line managers begin to take part in the conspiracy. People start to believe that they should continue to be rewarded for working safely and, when the prizes disappear, where does the motivation to work safely go? Out of the window.

If we are going to spend money and energy in improving safety, why not do it in a way that creates *sustainable* improvement? This way we only have to put in that energy burst *once* to make the improvement, needing much lower energy input to hold the gains (Figure 2.2).

The answer to sustainable improvement lies in the organizational culture itself – the climate in which people operate; the values and beliefs that the organization declares; the application of those values and beliefs by managers at all levels; the feeling amongst everyone that 'this is the way we do things round here'.

Is cultural change easy? Of course not – which is why many organizations get stuck on their individual plateau, and waste unnecessary energy trying to move off it.

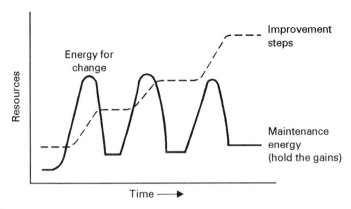

Figure 2.2

In this chapter, we will explore a simple strategy for sustainable change – sustainable improvement in safety performance:

Stage 1: assessing your current situation
Stage 2: deciding where you want to be
Stage 3: key issues within the change process
Stage 4: setting strategy and targets
Stage 5: laying plans and resourcing them
Stage 6: doing and reviewing.

This is a structure you can use wherever you are currently positioned – whether you need a radical restructuring of your approach, or whether all you need is a little fine-tuning of an already good approach and performance. The first step is understanding the position from which you are starting – your performance and its characteristics, your strengths and your weaknesses. The next stage is deciding how much better you want to be, what the characteristics of that new, desired state are, and the implications of bridging the gap – both in terms of resourcing and of the necessary change in organizational culture. The next stage is really one of understanding the issues relating to the change you will have to go through, and those factors that help and hinder the change process.

With an understanding of the size and nature of the gap to be bridged and ways to help you do that, the next stage is to set out a strategy for the change, with targets that will help you along the way. With strategy in place the next stage is the preparation of plans of work that will deliver the strategy, with a clear view of what resources are needed for the successful outcome of the plans and, finally, action itself – and the means of tracking and sustaining progress through structured reviews.

Quite a journey! Let's start with the first step.

Stage 1 Assessing your current situation

Before embarking on any change, it is helpful to know where you are starting from. This may seem a bit obvious, but we commonly take a lot for granted, and we need to take account of strengths as well as weaknesses in our arrangements for managing safety. Another important feature of change is understanding the gap between the current and desired position, to enable the journey of improvement to be planned.

There are all sorts of ways of assessing your current situation. A simple and common approach is to use measures of your current safety performance. Have a go at listing the measures of *your* current organizational safety performance.

The measures of my current safety performance are:

-
-
-
-
-

What did you get? You should at least know what your accident rates are – are they good, or bad, or average for your type of operation? How do you know? How much are your accidents costing you? What occupational ill-health problems do you have? What environmental problems are you facing? Are your operations causing problems for your local communities? What implications might all this have on your business? Are accidents causing delays in the supply of your product or service to your customers? What impact is this having on your business? How is your safety performance affecting employee morale? What is that costing you in terms of employee motivation and productivity?

Some of these questions can be difficult to answer in quantitative terms, and it is helpful to use some sort of structured assessment by which you may take a reasonably quantitative view of your performance which is not solely focused on accident statistics (since these are a very limited means of measuring overall performance – especially for tracking improvement in performance). If we look upon our safety management arrangements in terms of processes (Figure 2.3), it becomes more obvious that the output performance measures are dependent on the processes that deliver the

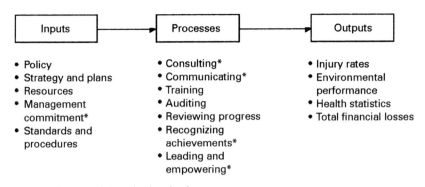

* Key indicators of organizational culture

Figure 2.3

outputs, and on the inputs into the management processes themselves. Any assessment of your current position, therefore, should take into account the outputs, the processes and the inputs.

There are some pretty detailed, commercially available assessment models which you can use to position yourself, and there are organizations that will, for a price, carry out the assessment for you. Alternatively, you can construct a fairly simple model for yourself, providing you are confident that your assessment model is comprehensive, and can be used to measure progress as you implement your improvement plan.

The basic components of an assessment model should include the following elements, which are fundamental to *any* successful safety improvement strategy:

● policy
● strategy and plan
● standards and procedures
● training
● auditing
● organizational culture
● resources
● results.

Each of these elements may be broken down into a further level of detail which picks out the key conditions and activities that together would characterize the best imaginable state for the element. You can attach 'scores' to each of these conditions and activities to position your current state against the best imaginable. The detail may, in a relatively simple form, look something like the framework which follows, each of the conditions and activities expressing the best imaginable state, into which a 'scoring' system could be grafted.

Policy

1. Signed by the most senior manager in the organization.
2. Defines concisely the scope of its application (e.g. which part of the organization, which geographical locations, and that it covers not just safety, but also health and environmental matters).
3. Contains a statement of commitment to operating in compliance with relevant safety, health and environmental legislation and company safety, health and environmental standards.
4. Defines the responsibilities of:
   - the most senior manager
   - people with specific (e.g. legal) responsibilities
   - line managers
   - all staff.
5. Defines arrangements for making the policy effective.
6. Contains an organization chart showing lines of accountability.
7. Contains short statements of principle to the effect that all accidents are preventable, and that safety is valued equally with other key business activities.
8. Defines arrangements for consulting with staff.
9. States who is responsible for the policy (should be the most senior manager) and how frequently it is updated (the policy itself should be within its valid date).
10. All employees are able to state these short statements of principle, and what their individual responsibilities are.

Strategy and plan

1. The strategy contains all eight elements listed earlier.
2. The strategy states, for each element, what will be achieved.
3. The prime responsibilities for achieving the strategy are clearly stated.
4. The strategy statements have the capability to deliver all the statements of intent in the policy.
5. The plan contains all eight elements listed earlier and has the capability to deliver the strategy in full.
6. The plan has a clear timetable and milestones for each of its elements.
7. The responsibility and lines of accountability for delivering each part of the plan are clearly stated.
8. The success criteria for each part of the plan are clearly stated.
9. The resources for delivering each part of the plan are identified.
10. The plan has been subject to consultation with, and contributions by, employees.

Standards and procedures

1. All the relevant organizational safety, health and environmental standards have been identified.
2. There are arrangements for identifying the needs for local procedures, as work changes.
3. There are procedures for all the relevant standards.
4. The procedures are up to date (they reflect current practice, roles and responsibilities).
5. The procedures are written in simple language and clearly identify the required standards to be achieved, as an aid to auditing.
6. When issued, each procedure is communicated to those people to whom it is relevant.
7. There is a system for evaluating and reviewing the procedures.
8. A named person is responsible for reviewing and updating all the procedures.
9. Employees affected by the procedures have an input into writing them.
10. Where relevant, specific responsibilities are made clear in each procedure.

Training

1. Safety training relevant to all types of job is identified and listed.
2. Safety training specific to different types of jobs is identified and listed, and used for individuals' training programmes.
3. Essential procedures are discussed with new starters at the time of joining.
4. Safety training is given priority – failure to attend a safety training session is a rare event.
5. There is a person responsible for ensuring that safety training programmes remain relevant and up to date.
6. Training courses are professionally designed.
7. Trainers are selected and trained to ensure they are capable of training others effectively.
8. Team meetings are used to discuss relevant safety procedures to raise levels of awareness.
9. Records of individuals' training are kept.
10. Critical tasks are identified, given priority for training, and the effectiveness of the training tested by evaluating the understanding of trainees.

Auditing

1. There is a procedure describing the arrangements for SHE (safety, health and environmental) auditing.
2. Auditing is carried out on at least two levels:
   - local 'operational' auditing (checks on operational conformance to procedures)
   - independent 'specialist' auditing (checking that existing procedures are still relevant and conform with current legislation and company safety standards, and reviewing the effectiveness of operational auditing).
3. There are programmes for both operational and specialist auditing.
4. A single person is responsible for the design of the audit programmes.
5. There is clear responsibility for carrying out the operational auditing programme at local level.
6. Everyone is involved in operational auditing.
7. Auditors are trained in the essential techniques of auditing.
8. Local arrangements for ensuring that actions to deal with non-conformances are assigned and followed through to completion.
9. There is a programme of specialist auditing to cover all the safety standards and procedures, with competent specialists assigned, and with each specialist audit being subject to formal reporting with actions being assigned and followed through to completion.
10. Performance to the overall audit plan is reviewed at the highest level.

Organizational culture

1. There is a clear statement of organizational values emphasizing the primacy of safety, health and environmental performance, and the personal behaviours that are expected.
2. Unsafe behaviours are never left unchallenged.
3. Managers demonstrate high personal standards of safe behaviour.
4. Recognition is given for performing to safety, health and environmental standards at all levels in the organization.
5. When an accident occurs, the emphasis is on learning and improving, not on allocating blame.
6. Measurement of progress is based on 'input' and 'process' measures, not just on 'output' measures.
7. Managers willingly receive feedback from their staff on their performance in managing safety, health and environmental issues.
8. Managers give priority to safety issues at team meetings.
9. Managers spend time in the workplace talking with staff about the safety of their operations.

10. Everyone is aware of their team / department / business safety performance, and can talk knowledgeably about what is being done to improve
performance and what they are doing in support of the overall
improvement effort.

Resources and results

1. Prime resources are assessed through careful analysis of the tasks
involved.
2. Assessment of resources includes both expenditure costs and
manpower.
3. Allocation of manpower takes account of the current workoad of those
involved.
4. Prime resources are budgeted and allocated
5. Resource allocation is approved at a management level senior enough to
deliver.
6. People from all levels in the organization are involved in resourcing the
plan.
7. Results are based on progress to plan as well as output measures.
8. Results are widely communicated and publicized.
9. Achievements to significant milestones are publicized and celebrated.
10. People are recognized for their contributions to the successful delivery
of the plan.

This structured approach to assessment gives you a pretty objective view of
your areas of strength and weakness, and is a baseline against which you
may judge your progress in your journey of improvement.

Stage 2 Deciding where you want to be

Having established your current position, the next step is to decide where
you want to be – what sort of safety performance do you want to achieve?
But even before you generate this vision, you should explore your values
with respect to safety. What values, or beliefs, do you hold about safety?
Values are the bedrock against which you can judge the alignment of your
vision, your policies, your strategies, your plans and your progress in
achieving the change in systems, culture and behaviours that will deliver
your vision.

These are some of the values commonly held by the best performers.

'All injuries and work-related illnesses can be prevented'

*'Nothing is more important than safety . . . not production, not sales,
not profits'*

> *'Safety is a management responsibility ... and an individual responsibility'*
>
> *'Safety is a way of life ... around the clock'*
>
> *'Safety is a condition of employment'*

In testing out your corporate values, being honest and searching is essential. It is not enough to copy fine sounding values in which you do not really believe. Statements of values should be the product of personally held beliefs, and should be shared by those at the top leading the organization. However hard you try, if your statement of values does not line up with your personal beliefs, your personal behaviour will not be consistent with the stated values – and it will show.

Having established your corporate values, you are well placed to establish your vision, and there are techniques you can use to help you do this. Two simple approaches are:

- brainstorming your desired state
- benchmarking yourself against other organizations whose safety performance you would like to emulate.

Let's take each in turn.

Brainstorming your desired state

Brainstorming is a simple technique used to generate ideas and views which can then be structured to provide a focus for action. It may be used to generate organizational values, ideas for improvement, problem solving, and more. The simple rules for brainstorming are:

- have an agreed objective
- take five minutes for individual reflection
- collect ideas (write them down) as they are expressed by individuals – no interpretation!
- no constraints on ideas or views – even the most way-out ideas!
- no evaluating ideas or views during the brainstorming process
- time for generating ideas limited to around 15 minutes.

Once all the ideas are out and written down, they may be clarified and grouped to help focus on primary issues or actions. This is the stage at which open discussion is needed to gain concensus, and to structure the output into actionable form.

Examples of questions which may be brainstormed to develop a vision of your desired state of safety management are:

'If we were really excellent at managing safety:

- what would our management systems look like?
- what would staff be saying and doing?
- what would managers be saying and doing?
- what would people outside the organization be saying about us?
- what would characterize our performance?
- what would be the advantages/payback? '

Examples of typical output responses to these questions might be:

- we would have minimal financial losses as a result of accidents
- we would have a clear safety improvement plan containing the essential elements, with responsibilities, milestones, targets and measures
- staff would be able to describe the safety improvement plan and what their individual contributions were
- staff would be talking with each other about workplace safety issues, and how to improve operational safety
- managers would be talking with their people in the workplace about the safety of workplace operations
- managers would not pass by an unsafe condition or behaviour without taking action
- managers would be seeking to identify the root causes of accidents, not seeking to apportion blame
- other organizations would be visiting *us* to see how we achieved such a good safety performance
- we would be getting external awards from safety institutions
- our injury frequency rates would be down to. . .
- our conformance to environmental legislative consents would be 100 per cent
- we would have no occupational health-related cases
- working days lost as a result of injuries would be down to. . .

. . .and so on.

If you are able to express your desired state in these sorts of terms, you can begin to set out some targets for your change process, both in numerical output terms (injury rates, environmental performance, etc.), in cultural terms, and in terms of management systems. Comparing these targets against where you are currently allows you to assess the gap you are going to have to close.

Another, supportive approach is to ask yourself: 'If we were really excellent at managing safety what would we:

- see or hear more of
- see or hear less of '.

The answers you get to these questions can help you to see which of the characteristics of your current safety-related activities could be built on (your strengths) and which are most in need of change (your priorities for improvement).

These approaches, held in an open and constructively critical way, allow people to express personal views, beliefs and ideas that would otherwise be difficult to get out into the open. Brainstorming these issues can be carried out at any level, allowing you to gather views from all sectors of your company staff. Not only can it be fun, but it gives you a lot of information about people's views in a relatively short space of time. It is also a very simple way of getting people involved in the change process – particularly managers, who will have to lead it.

There are slightly more sophisticated techniques for involving people in resolving problem issues in a way that opens up the problem, identifying the barriers to change and progress, and the solutions to overcome the barriers. They can be a very powerful way of involving people in generating their own solutions to problems, frequently gaining an exceptional degree of ownership and action. They are especially useful for making progress on issues where progress has been particularly difficult to achieve – often where traditional ideas of working have to be overcome to make any real progress.

Benchmarking

Questions that always arise when companies start to challenge their safety performance are:

'how much better do we *want* to be?'
'how much better *can* we be?'

Ideally, of course, we would like to achieve *zero* losses to the business through accidents, but since most companies are so far from achieving this, it is usually only useful as a long term vision. At the same time, setting limited improvement targets can be under-ambitious.

A practical way of raising realistic expectations for improvement is by comparing yourself against other companies which are either in the same business sector as your own, or which have similar operational hazards, *and* which are known to be leaders in the field of safety management. At the simplest level, you can use this approach to compare your output performance measures (injury rates and other loss statistics) with those of others. This will tell you how much better other companies with similar activities to yourself are at managing safety, and will let you know the sort of accident rates to which you might aspire.

At the more detailed (and useful) level, however, benchmarking can tell you much more – not just the *what* (the performance), but the *how* (the practice). Companies which have a significantly better safety performance than your own will almost inevitably have already trodden a similar path that you are preparing to take. In doing so, they will have learned much along the way that can be useful to you – those things which you can positively and practically do to help your progress, and those things which are worth avoiding. It is always worth learning from others' mistakes (and, of course, their successes) before you start.

Although you will always want to select companies which have a significantly better performance than yourself in order to learn as much as possible, you should not confine yourself to benchmarking against companies similar to your own. Similar organizations may be able to tell you a lot about the ways in which you might approach specific issues common to your organization (e.g. hazard studying and risk assessment on petrochemical plants) but different types of organization can give you good ideas in those many common areas of safety management: training and auditing are examples.

When approaching other organizations, be prepared to share with them your objectives and plans, and any learning you have gained so far; they may be ahead of you, but an exchange of ideas may also generate some valuable ideas for them. It's worth keeping in touch with those organizations with which you have positive exchanges. In time, they may be able to learn from you and, as they continue to improve, you can continue to learn from them.

To make the most out of benchmarking, treat it as a continuous tool to help you 'leapfrog' the natural learning process, and therefore accelerate your rate of improvement. Fortunately, SHE matters are not regarded so much as a business competitive issue, so that benchmarking on these issues is more of a co-operative process than a competitive one. Benchmarking with the best – or at least those a good deal better – will help to raise your expectations as to *what* you can achieve, and often a lot about *how* you might do it. It also makes you aware that you are not alone in the search for improvement, prevents your internal perspective from becoming too narrow, and gives you external 'partners' from whom you can draw mutual encouragement.

Stage 3 Key issues for the change process

So far, we have covered the assessment of your current position, and the ways in which you can decide how much better you want to be. By doing this, you will know in very broad terms what you want to achieve, and some feel for the size of the task ahead – in terms of output measures, cultural change and safety management systems.

It would be natural at this stage to move into the 'planning and doing' phase. But before we do, it's worth spending a little time thinking about some key issues which will affect the change process, and therefore the rate of progress of the changes you will need to make. Important *people* issues are always involved in the process of change, and the attitude of people to the change profoundly affects its success.

The key features in successful change are management leadership style and management technique in gaining acceptance of, and involvement in, the change. We will take the two in turn.

Management leadership style

The first point to remember is that on the whole people like to *do* things (feel in control) rather than have things *done* to them (feel they are being controlled). Change is something that everyone feels personally about. If we don't feel a change is justified, or if we simply don't understand it, we naturally resist it. We will accept change if we feel it is justified and not contrary to our interest. We will accept *and* support change if we believe it is in our interest. And we will usually accept, support and actively promote change if we believe it is in out interest and are made to feel an important part of the change process. So, if I am to be a successful part of someone else's change:

- I have to feel it is justified
- I have to believe it is in my interest
- I have to be committed to playing an active part in it.

Getting this level of buy-in can be difficult. Even if there are no obvious disadvantages to me, what are the hidden effects? How will *I* have to change? What happens if I find the change difficult? Will there be penalties?

Change cannot effectively be managed. It has to be led. And, at its best, that leadership embraces values, vision, determination, exemplary behaviour and an understanding of the personal transition process that people need to go through to make the change successful. The most effective leadership is underpinned by strong, clear values.

Earlier in this chapter, we explored values and vision, and the importance of the most senior managers in the organization generating and believing in them. This then becomes the driving force for the change, which expresses itself in senior management determination to see the change through. This determination, clearly communicated, will rapidly be reflected further down the management line, provided that the behaviour of the change leaders exemplifies the values underpinning the drive for change. Leading by example is a crucial aspect of leadership behaviour. If, as a manager, I tell my team that it is important that we all take action on important unsafe situations, and I pass by such a situation myself, what does that tell my team about my underlying values? I immediately lose my credibility as a leader.

Understand the change process itself

Earlier, we touched on some of the barriers to change. Let's take a look at this in a little more detail, and at the stages in the change process.

When people are faced with a change, their first reaction may well be to regard it as a threat or a burden – '. . . this is going to mean that *I* will have to change . . .' or '. . . yet *another* initiative that will cause me more work . . .'. This attitude creates a feeling of rejection of the change (see Figure 2.4).

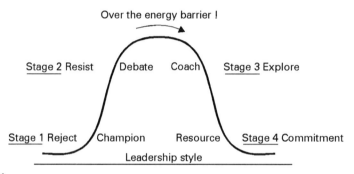

Figure 2.4

This is often the most difficult stage for the change leader. The challenge here is to understand the attitudes and concerns underlying the rejection, and to be able to start to meet those concerns with the positive aspects of the change. Here the leader is not only acting as a champion for the change, but is exploring and meeting the main concerns openly and head on. This is the stage where assertiveness may be necessary to move on to the next stage.

Once people start to engage in open debate about the change, they are moving on to the next stage – resisitance. This is where people start to see the need for change, but are not convinced that it is in their best interests, or fear the 'unknowns' which the change may bring. They resist a move from the comfort and security of the current state. Here, the leader needs to listen empathetically to people's concerns, responding to those which can be answered directly and capturing those which are more difficult to deal with, for considering and answering later. A common problem for people at this stage is belief (or lack of it) in management 'say and do'. For example, if in the past safety has taken second place to getting the job done quickly, and management is now saying that work will only be done if it has been properly assessed and adequate safeguards put in place, and if I then follow the new principles and the job is slower as a result, will I be criticized or will I be valued for following the new principles? Certainly if, in the past, I have experienced managers saying one thing and judging me on another, I will now have good reason to be sceptical.

Open and honest dialogue, and a listening style, is important in this stage. Allowing people to express their concerns and any previous bad experiences, and to test out their perceived consequences of the change, helps them to work through their concerns and anxieties. It is also essential to keep an open mind here – not about the direction of the change, but about its shape and pace. This is the stage where people start to become more accepting of the change if they are able to exert some influence which helps to meet their main concerns, particularly if there are common and strongly held concerns across the team or organization. Problem-solving techniques may be used to deal with particularly knotty issues.

This process of open dialogue is an important step for people in starting to come to terms with the new situation, helping to loosen the ties with the current state and releasing them to engage with the practical issues of the new situation.

Once people have had the opportunity to debate fully the purpose and consequences of the change, they are ready to start exploring the change, and their part in it more confidently. This is the stage which people need to take at their own pace, and not be overwhelmed with detail. They need to see practical ways through the change with which they can feel comfortable. Teamworking can be a big help here, people testing out ideas interactively and giving mutual support where problems or uncertainties arise. During this stage the change leader acts as a coach – not pushing, but steering and guiding, helping the team and individuals to think through options, practical steps and potential problems. As people work together on the way forward, the team often starts to feel increasingly confident and in control of the new situation and much creative energy can be generated.

When the team has achieved a practical sense of direction, it is ready to move forward into the final stage – commitment and action, where the role of the change leader is to stand back and allow the team to make progress, encouraging them and recognizing their achievements. Careful steerage is sometimes needed to ensure the team remains on track and focused on the goal. Another helpful role for the change leader in this stage is to get the team to look backward from time to time to see how far they have come, and to help them to see what they have learned from the transition process. These pauses for reflection and learning make subsequent changes easier for the team to manage.

Throughout the change process, and beyond into implementation towards the goal, an additional important role for the leader is constantly to reinforce those behaviours which are supportive of the direction of change, of the teamwork spirit, of the ultimate goal, and, of course, careful counselling of counter-productive behaviours when they arise. This reinforcement acts as an additional catalyst to progress.

This whole process of feedback to help reinforce positive behaviours is effective not only down the line, but up the line. Leaders, like everyone else, have their strengths and their weaknesses and, again like anyone else, need feedback on their performance. Naturally, they should get this from their own line manager, but it is equally helpful to receive 'upward' feedback from their team, especially since their leadership behaviour affects the team more than anyone. This mutually reinforcing and supportive feedback is a feature of the more enlightened organizations.

Managers are often nervous about introducing upward feedback, and an open uncritical and trusting climate needs to exist before it is likely to be accepted. With these provisions, however, upward feedback can be a significant accelerator to progress, and there are ways of conducting upward feedback anonymously so that people do not feel constrained in giving constructive criticism.

Even more progressive approaches to performance feedback are '360 degree' feedback, i.e. feedback from

- your manager
- your peers
- the people you manage
- customers.

All these forms of feedback may be used on an individual or on a team basis and, constructively and openly applied, helps an organization to learn rapidly from itself and accelerate its progress towards its goal.

Gaining acceptance and involvement

We have explored the first of the two key factors in successful change – management leadership style – in some length, and along the way we have touched on some of the aspects of gaining acceptance of the change process. We will now explore the second of these two factors – gaining acceptance and involvement – in a little more detail, examining some tools and techniques which can help the change process along.

Involvement in visioning and problem solving

Earlier we discussed the importance of the most senior management generating a vision of where they wanted to be, and how this technique helps to create a sense of direction, goal and purpose – and ownership of them. This technique is no less powerful for generating a sense of dissatisfaction with the current state and a sense of direction and purpose amongst staff further down the organization. It is surprising how much commonality in view there can be between widely different levels in the organization when people have the opportunity to explore their basic values about safety and their perceptions of the scale of the improvement that could be made given the right level of organizational commitment.

The simple act of generating this vision not only enables the change process – since it allows people to lead themselves through the first three stages of the change model (Figure 2.4) – but tends to generate some improvement activity spontaneously (stage 4 of Figure 2.4). However, the technique needs some careful facilitating to make sure the exploration process is carried out in a way which allows everyone to express to their satisfaction their own personal views, and which steers a logical course from opening up the issues to closing down on to solutions.

Acceptance of new ideas

Earlier we examined the change model and the needs of people to move through the four stages. Whilst the change journey is pretty much the same for everyone, the pace of the journey varies from person to person (see Figure 2.5). There will always be a small number of early enthusiasts, some of whom may act as helpful catalysers for change, others who are quickly motivated by new ideas, but whose enthusiasm may be short-lived. The change process really starts to get underway when the early leaders begin to emerge; these are the people that consider the merits of the change carefully, and who take positive action once they are convinced of the benefits. As the early leaders gain momentum, others start to follow – some sooner, some later – and eventually those most resistant to the change follow on.

It is worth knowing who the early leaders are likely to be and paying special attention to them in the early stages of the change process, giving

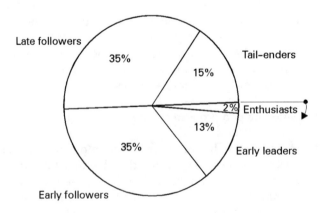

Figure 2.5

them encouragement without forcing the pace of acceptance. Logical and carefully presented arguments, and a display of personal commitment to the goal and the change process is likely to be most influential with these early leaders. As they begin to gain acceptance, they can be helpful by influencing the acceptance of others and may be prepared to take on a more proactive role in the change if encouraged.

The change equation

Earlier I mentioned generation of dissatisfaction with the current state and a vision of the future state to help with generating a more positive view of change in people's minds. These are two factors in a now commonly used model called the 'change equation', which applies equally at personal, team or organizational level. The equation:

$$D \times P \times V > C$$

says that for change to occur, the combination of dissatisfaction with the current state (D), vision of the future state (V) and practical steps to get there (P) has to be greater than the perceived cost of the change (C), whether that cost is material or emotional. If the perception is that the cost outweighs the other three factors – whatever the reality of the situation – the change will not occur. If the three change factors are enough to make people believe that it is in their interest to move from the present state (dissatisfaction), and if they have a clear view of the changed state and how it will serve their interests (vision), and if they can see clearly how the transition can be achieved (practical steps), then the change is much more likely to take place.

Preparing for change

When decisions about change are being made at the top of the organization, it is easy to lose sight of the impact of the announcement of the change further down the organization. People, especially those that will have to lead the change process, do not respond well to having the change 'dumped' on them. It is important, therefore, to think ahead about how to break the news of the change, and be mindful of how this may be done in a way which minimizes the work which change leaders at the local level will have to go through to help people through stages 1 and 2 of the change model.

Consider at which stages in the process of generating your values, vision, strategy and plans you might engage in communication and listening to feedback. Consider how you are going to make managers (your change leaders) feel in control of the change process by keeping them informed and involved. Consider the questions and concerns that will meet the announcement of the change, and prepare responses which managers can use as a basis for answering questions.

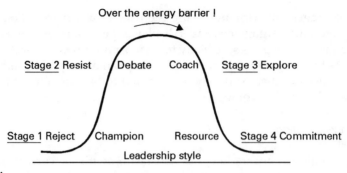

Figure 2.6

Consider ways of exposing the change to people in a way that doesn't leave them overwhelmed and disconnected from the decision making. Exposing the change in easy stages allows people to come to terms with each stage and makes them feel that the entire detail of the change hasn't been pre-ordained. For example, when your most senior management team has agreed its values and vision, let people know in a very up-front and confident way. Making the values and vision very public and prominent shows your determination and commitment to them, helping their acceptance further down the organization. Briefing line managers in advance of the announcement recognizes their special role as leaders of the change.

At this stage it actually doesn't matter if you haven't agreed a strategy. Let people know that this is the next stage of the process – that you wanted to share the early decision about future direction with them, that you haven't got all the answers yet and that if they have any ideas they want to contribute, they will be welcome. Then, when the strategy is developed, you can communicate again, letting people know what the next step will be, and inviting feedback.

A little early thought in how the change should be introduced is always a sound investment. It helps to minimize suspicion and resistance to the change, makes the job of the change leaders easier, and helps to ensure early acceptance of the change so that progress can be made quickly once the practical implementation of the change has been worked out.

The role of communication

Two-way communication is a vital part of change, with both formal and informal communication playing a part. There are many methods of communication, but the two most important ones are remote and face to face. Both have their uses. Remote communication (letters, memos, electronic messaging, notice boards, etc.) is fine for getting key information out to people in a consistent way which ensures they all get the information at the same time. It is also fine for newsletters, which can help to reinforce key information in an easy-to-read form. However, during a change period, these should be regarded as supporting communicating processes, simply because they are one-way.

During change, people need more information than usual, simply because of the additional and unusual events that are taking place, and need to talk about how they feel about the changes that are taking place. The only way to meet these needs effectively is face to face. This two-way communication:

- clarifies understanding
- allows concerns to be raised and discussed
- allows views to be exchanged and explored
- gives an opportunity to collect ideas.

Don't forget that if *you* don't do the communicating, the 'informal' communication system – the 'grapevine' – will do it for you. When this happens, you effectively lose control of events, people feel neglected and isolated from the change process, and you stand the risk of misleading information being passed on. Control over the change process depends on good leadership, of which communication is a key element.

Managers often feel vulnerable about face to face communication with their teams because they may be unsure themselves how things will turn out, and because they think their teams will expect them to have all the answers. Relax. Not knowing all the answers is a positive *advantage*, for two main reasons. First, no reasonable human being can possibly have all the answers in a change situation. Second, not having all the answers allows people to influence and contribute.

Here are some basic pointers for face to face communication.

Be honest

People like to be told the truth. Tell them honestly what you understand about the change, and about uncertainties and options. Whilst the decision to make the change will not be up for debate, the means of making the change (the 'how to') will often be open to influence. People will usually respond well to the 'warts and all' approach, since it tells them that uncertainties and problems are not being hidden, and that they have an opportunity to influence the path of the change.

Be prepared

Managers have much more confidence about engaging in face to face communication if they have thought through how they should handle the process. A simple structure for a communication session may look something like this:

- explain the change
  - say why the change is needed, what the expected outcome of the change is, and what the perceived benefits are
- describe its impact
  - as best you can with the information you have, explain what the change will mean in practical terms, both for the organization as a whole and for the team
  - what you know of the way in which it is proposed to make the change (or what the options are)
  - any knowledge you have about similar changes having been successfully made elsewhere
- ask for feedback
  - be prepared for questions like 'what will this mean for me/us?', and 'what are the hidden agendas?'
  - allow people free voice for their opinions
- discuss the feedback
  - answer questions and views openly and honestly, and commit yourself to resolving reasonable questions that cannot be answered at the session
  - collect views, and any ideas, for feeding back up the line
- look for support and involvement
- test the team for its support, asking for practical ideas which might help implementation, and explore how the team might contribute their individual knowledge and expertise.

Listen!

Face to face communication is only effective if it is truly two-way, which means that managers have to *listen* to people's concerns and opinions. Listening means understanding and accepting what has been said – both what is expressed and what emotions and concerns lie behind it. It is helpful to play back what you think has been expressed. This enables you to be sure you have a clear understanding of the point made, and it tells the person that you are taking the trouble to understand. Where you suspect that a question or concern conceals some other issue, it can be helpful to ask some questions of clarification, which might draw out some other issues. Both these approaches help to make sure that you fully understand the issues that are raised – and some of the less directly expressed views and concerns – and that people feel you are actively listening to those concerns and needs.

Involvement through empowerment

The enormous leverage that can be given to a change process from people being committed and involved is almost always underestimated. Let's look at that change process model again (see Figure 2.6, p. 36).

In the light of what we have explored about communication and involvement and other criteria for successful change, let us look at this process in a slightly different way, more in terms of how people are feeling about their part in the process.

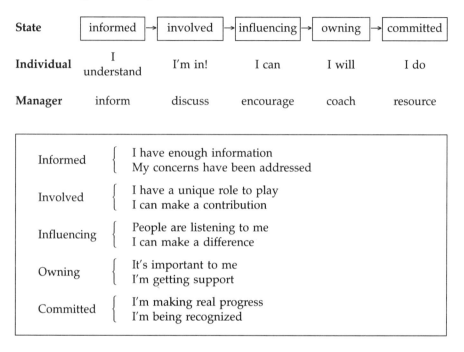

This path through understanding to involvement to commitment is all about giving people the encouragement, freedom and resources to believe they *personally* can make a real difference. This feeling of empowerment gives enormous acceleration to the final stages of the process – ownership and commitment. The best contributions a manager can make to this process is not to manage, but to *lead*, and provide the information, encouragement, coaching and resources that people need to be able to do the job with both responsibility and freedom, and to dismantle barriers that restrain progress. This approach, which often requires managers to undergo a philosophical shift in their role, can give organizations a remarkable edge – not just in their management of change, but in the effectiveness of their business as a whole.

Stages 4 and 5 Targets, strategy, plans and resources

Achieving significant improvement in SHE (safety, health and environmental) performance is a big task. It is no different from any other major business change, however, and its planning and resourcing just as necessary. No major change – especially one which will take place over a period of several years – will be sustained unless it is well planned and resourced.

Mapping the journey

In setting out on this journey of change and improvement, it helps to have a structure, especially where the change is large and complex. A typical path might look like Figure 2.7. It starts with the *vision* that we explored earlier in this chapter – your generation of a conception of the characteristics of your safety management systems, culture and performance in your desired state.

Setting this vision against a knowledge of your current state gives you a concept of how much change will be needed and, as a result, the scale of the task ahead.

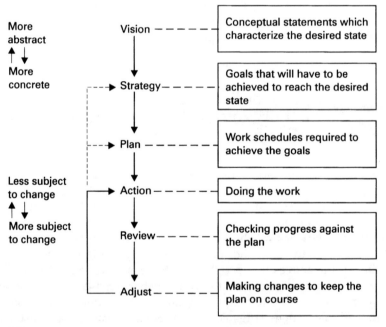

Figure 2.7

The next stage is setting your strategy. This entails your describing clearly the goals that will need to be set to reach the desired state. Here you are starting to get into a clearer structure, identifying the key elements which comprise the change and what they all look like to achieve the change successfully. These key elements may be easy to identify from your vision statements, but it is always worth going back to a structured model, for example the process model in Figure 2.3, to make sure you have identified the key change elements.

For some elements (e.g. culture), you may decide that the gaps between where you are and where you intend to be is so wide that you need to take the change in easy stages and plan for each stage in turn. It is worth looking for those stages across elements which are naturally reinforcing or parallel, to decide whether they should be developed together within the overall plan.

This 'whole plan' approach may seem detailed, but it will help you to see the totality of the task ahead. This enables you to:

- assess the resources you will need to implement the plan
- construct a programme which is a realistic match for the resources available
- set the programme elements and their respective work packages into logical order
- identify priorities and allocate resources accordingly
- monitor overall progress
- avoid overloading individuals with too much of the plan's implementation at any one time.

The strategy should contain statements of intent about each of the key change elements. For *policy*, for example, the strategy statement may be 'There shall be a statement of safety policy which clearly expresses the company values and commitment to safety, which describes the means and responsibilities through which the commitment is translated into practice, and which is understood by everybody'.

Having established strategy, the next stage is planning. Here we are getting into the detailed 'how to', often characterized by *what, how, who* and *when*.

The *what* are the stages of the task. For policy, for example, the stages might be:

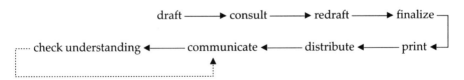

This is a fairly simple task. Others – for example planning the implementation of a training strategy for safety – would be more complex. In this case it is helpful to have an overview, a simple plan, with each stage in the plan subject to a sub-plan with a further level of detail. This building up of the picture avoids getting tied up in too much detail in the early structuring of the plan. All this may seem pretty obvious, but unless the plan is mapped out, it is easy to miss out important stages, especially when you get into the 'doing' of the plan. In the 'policy' example above it would be easy, for example, to miss the consultation and communication stages as you focus on getting the policy right.

The *how* prompts you to explore the options for achieving the plan elements. In the policy above, for example, the drafting stage is straightforward, but there will be actions about how to go about the 'consult' stage, and you would be looking to choose an option which is both effective and economic. The 'how' is an important part of the plan becuase it determines the resources that you need to allocate to see it through. So it is worth spending some time exploring the options, to choose that which is a good balance between simplicity, economy and effectiveness.

Knowing how the plan will be implemented gives you a good idea of the skills and resources needed to carry it through successfully, helping you to decide *who* should be assigned to the task. If you do not have the skills, or the resources, in-house you may need to buy them in, but doing as much of the implementation of the plan as you can in-house gains much more ownership; it is usually surprising how much latent talent there is around. As well as choosing those with the right skills, it is important to choose people who are also committed to the task. Enthusiatic volunteers with the right skills are the best possible choice!

Finally, the timetable – *when* to carry out and complete the task. Again, this is made easier by having established the 'how' and the availability of those chosen for the task. Deadlines should be stretching, but realistic. Setting deadlines which are too tight can be frustrating, especially when a deadline is compromised by unforeseen complications. Setting very loose deadlines can result in complacency and loss of drive. An effective approach is to treat each part of the plan as a formal project, and manage it as such. We will take a look at this a little later, in 'doing the work', which is the next stage of the model in Figure 2.7 – 'action'.

Stage 6 Doing and reviewing

Very simple plans will not usually require any review until they are complete. Plans which are more complex and time consuming, however, need to be reviewed periodically to check on progress – to see if the work is progressing both to schedule and within the allocated resources, and to programme. It is a rare programme that runs without any hitches, especially where the ground being covered is new. Unforeseen problems can arise, new ideas emerge and experience is gained, and new demands on people's time conflict with their commitment to the task. All of these influences are quite natural, but can have a profound effect on the programme. *Reviews* of progress are essential to make sure that these constraints – and opportunities from new ideas – are not left unmanaged.

Reviews may have different outcomes. One outcome is that progress is fine and no action is necessary. Another outcome is that there are deviations from the plan that can be managed by simply allocating more resources, or taking up slack in the programme, to get it back on course. Yet another outcome, where unforeseen problems are too great to contain within the project programme and allocated resources, is to revise the plan itself – a course of action that is preferable to allowing the project to drift out of programme or resources. Very occasionally, it may become apparent during the implementation of the plan that the underlying strategy is out of balance, and needs to be readjusted. Going back to strategy in this way should be an unusual event, but should be accepted if the eventual success of the plan is seriously compromised.

Monitoring the overall progress of the programme must be the responsiblility of the top level management, for two reasons. The first reason is that it gives credibility to the programme. The second reason is that this is the level at which significant decisions about resourcing and changes of direction need to be made, and from where encouragement and recognition for progress are most effective in sustaining the enthusiasm of those who are implementing the programme.

Give responsibility and accountability for delivery

It may be an obvious thing to say, but responsibility and accountability have to be given for delivering the programme. A practical approach is that responsibility for the programme as a whole is taken at a senior level, with responsibility for individual elements of the programme being delegated to others down the management chain. There may be further delegation of responsibility for the stages of each element, dependent on the scale and complexity of the task.

This delegated approach is fine, with the following provisions:

- that the person delegating retains accountability for delivery
- that those delegated to the task are given the time, resources and, if necessary, training to do the job effectively
- that the person delegating sets and carries out progress reviews with those doing the work.

Choosing the right people to work on the improvement process

Progress to the programme will depend very much on the enthusiasm and capability of the people implementing it. There two guiding principles worth bearing in mind: to choose the right leaders and to involve as many people as you can.

Choosing the right leader for each part of your improvement programme is just the same as choosing the right project manager for an important business project. Pick your most committed and capable managers to develop the key parts of the programme: if you can, let them choose the part in which they have a particular interest – that way, the commitment will be greatest. Give overt support to these leaders; make it known that the task they have taken on is important, and give them open credit for their achievements. You don't have to pick your most senior people: commitment and capability are the key driving forces, and it is an opportunity to develop some of the more promising middle or junior managers in your organization.

Give leaders the freedom to choose their own teams for the task, within broad guidelines for appropriate mix of skills and experience. Leaders will work much better with people they have chosen than those they have been allocated.

Involve the people

Most people are a source of surprising talent which is normally untapped at work. Once committed to a project, people will offer remarkable energy and time to making a project a success, especially if it is an added interest to their normal job. Where you can, make an open offer of involvement in the improvement projects, and see how your staff respond. If you are fortunate, your project leaders will have the happy task of choosing from the volunteers. If not, you will have to resort to selecting people to be involved.

Whichever route you take, make it known that their contribution will be vital. The more people you involve, the greater the speed of change and improvement, both because there are more people working on the improvement programme, and because there are more people being positively influenced by the simple act of involvement itself.

Doing the work: managing by projects

Earlier, we touched on the most time-consuming part of the improvement journey – implementing the plan. Typically, implementation of such a plan is a major task, and often spans a number of years. An effective way of dealing with a task of this scale is to break it down into manageable chunks – work packages – and treat each significant package of work as a formal project. The advantage of this approach is that most companies already have arrangements for project management, which give a formal structure to significant tasks. A typical project management approach is shown in Figure 2.8.

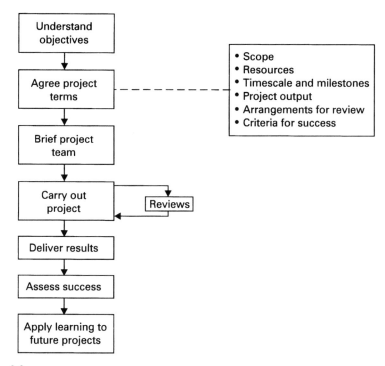

Figure 2.8

It is surprising how well the discipline of writing down and agreeing a simple project definition ensures a complete common understanding between the person responsible for the work (the project manager) and the sponsor (the customer) of the project, often saving unnecessary or misdirected effort. It also establishes criteria for the success of the project, and the means of reviewing progress.

Targets

So far, I have not mentioned targets, and you may be wondering why I have left it so late to introduce them.

There is nothing more designed to fail than targets without clear strategy and plans for meeting them. No company worth its salt would pluck next year's financial targets out of the air without some form of forecasting, based on a knowledge of the capabilities of the business within the context of the market in which it is operating. Similarly, targets for improvement in safety, health and environmental performance should not be set without an understanding of what has to be done to achieve the targets, based on an understanding of the required change.

Successful companies have clear, long term plans for development – plans which include growth, capital investment, market share, and so on. Improvement in safety, health and environmental performance should be on the same basis – a long-term (say five year) plan for *sustainable* improvement – one which is designed to hold the gains once they have been made.

Not only do targets have to be matched by practical plans to implement them, but by the resources to carry them out. It is easy to be over-ambitious in setting targets, especially where assumptions are made about the speed at which the plans can be realized, both in terms of the amount of effort needed to carry out the plan, and the available resources.

The pace of change is determined by the rate at which the necessary new management systems can be prepared and the rate of acceptance and implementation of the new systems.

Output targets – injury rates, for example – can only be indicative; they cannot be absolute. You can not set a target without understanding what has to be changed, and by how much. Conversely, setting out plans for improving key areas of your operations will inevitably result in output improvement – although you can rarely foresee how much. Targets, therefore, are a mechanism for driving improvement, rather than an absolute reference point for what can, or will, be achieved.

Another constraint on output targets is that they cannot be met by changing a single process or input. They are the product of many inputs and processes. The effect of these complex interactions cannot be predicted accurately in terms of outputs, What you *can* do, however, is to monitor the outputs to see how things are changing as a result of the input and process improvements you make.

Having made all these points, it has to be admitted that setting targets is almost irresistible. It is part of our culture. It is almost as though an inner voice is telling us that if we don't have output improvement targets, we will never drive ourselves to meet them. We'll leave it for now and return to targets and measures in Chapter 8.

Chapter 3
Starting with policy

'Policy is one of the most common illustrations of misplaced corporate optimism'

The purpose of a policy

What is the purpose of a policy? Well, with a few exceptions, it is a legal requirement to have a health and safety policy, so that is a good enough reason in itself. It would be straightforward to write a safety policy that would pass legal muster, but how useful would it be if that was all it was designed to do? All you may produce would be a relatively sterile document which has little real impact, and of little real added value to the safety improvement programme you will be preparing. So, what is the *real* value of a safety policy?

It is worth starting by considering what the fundamental values and aims of the company are, taking into account the requirements of all the stakeholders in your business. Why? Well, quite simply, a business succeeds or fails by the way it serves its stakeholders, and this is true from a safety point of view just as much as it is for any other aspect of the business.

Who are the business stakeholders? For every business there will always be three – suppliers, customers and employees – and for many businesses, two others – shareholders and the local communities in which they operate. Consider for a moment what the requirements of these five stakeholders might be in terms of the safety of your operations:

- suppliers . . .
- customers . . .
- employees . . .
- shareholders . . .
- local communities . . .

They might look something like this.

Suppliers

- Uninterrupted (no major incidents) business, which allows them consistency of supply
- Successful business which is growing, providing them with an increasing supply outlet.

Customers

- Reliable, intrinsically safe processes (no interrupted supply of product or service)
- Good product design which does the job well, can be used safely and has minimum environmental impact

Employees

- Safe operations
- Healthy working environment
- Effective training
- Good company image.

Shareholders

- Safe operation: uninterrupted business
- No costly litigation
- Environmentally sound
- Good company image

Local communities

- No dangerous incidents
- No health risks
- No adverse local environmental impact
- Information about hazards.

You can start to see from this list that there are common interests from one stakeholder to another – and, of course, employees can be shareholders and members of local communities. With an understanding of these interests, and with your own knowledge of the safety, health and environmental aspects of your operations, you are ready to start considering your policy.

Policy as the driver of detail

A good policy is the starting point for all the safety, health and environmental assurance activity that follows. As we will see in later chapters, it is not possible – or even desirable – to control every human activity at work by procedures, so there have to be some guiding values and principles that apply to everything that is done at work. It is worth giving serious consideration to these guiding values, because they will be the reference point for managers, for staff, and for all those who will be developing systems in support of your improvement plan.

Where do you start in determining your guiding values? Firstly, it shouldn't just be the product of a single individual, no matter how elevated. It should be the subject of serious debate at senior management level at the very least. There is a lot to be gained by opening up this debate more broadly. One group of your business stakeholders – the employees – will have some strong views about the values of its company, not least because they will be most affected by failures in the arrangements for managing safety. And, of course, the more views you collect, the easier it is to identify common, strongly held views which can be reflected in the values statement.

Scope, structure and content of a safety policy

Having expressed the organizational values as an opening statement of policy, a good policy document follows through with:

- the scope of its application
- the arrangements for the practical application of the policy statement
- key responsibilities for the practical application of the policy
- the arrangements for consultation and communication
- how the policy is to be reviewed and updated.

This sounds like a lot of detail, but in fact a policy only needs to contain as much detail as is necessary to be clear about how the values are translated into action, and how that action is progressed. So the policy doesn't need to contain, for example, all the details of your safety improvement plan; the policy simply has to reference the plan, what its review mechanisms are and who is responsible for it. The test of a good policy is whether you can follow conclusively – without breaks – how every detail of work done to sustain and improve safety performance traces back to the policy statement and, conversely, how the policy statement comprehensively translates into the necessary systems maintenance and improvement activity on the ground.

Scope of application

Defining the scope of a policy ensures that the boundaries of its application are clear. This need is fairly obvious for large corporations, in which the component businesses may need to write their own safety policies, for example, where they are limited companies in their own right. That a policy should state its scope of application may seem unnecessary for small, compact organizations, but you have to consider, for example:

- how the policy relates not just to your employed staff, but also to any contractors and visitors
- if you have operations on more than one site, whether it covers all the sites or just one of them
- if you are operating in a multi-occupancy building, how the boundaries of your responsibilities are defined in respect of (for example) emergency procedures.

Thinking carefully about the scope of application is, in any event, a useful reminder of what your business is really responsible for and, not unusually, identifies gaps in management responsibilities and sometimes in the provision of professional safety support.

Arrangements for the practical application of the policy statement

This is the part of the policy which describes how the values and statements of intent are translated into meaningful action. This is where the safety improvement plan fits in. As mentioned earlier, it isn't necessary to reproduce the plan itself in the policy. It *is* important, however, to be very clear about

- what the aim of the plan is
- what its essential components are
- how it is translated into local action
- what the direct and support responsibilities are
- who is responsible for ensuring overall progress to the plan
- the reviewing and reporting arrangements
- the means of keeping its profile high.

The aim of the plan is to sustain and improve on the current safety performance by developing those key elements of safety management that we explored in Chapter 2:

- standards and procedures
- training
- auditing
- culture
- resources.

A brief strategic statement for each of these elements, describing the intent for their development, and how progress is to be measured, is all that is necessary providing the detail in the plan is clearly consistent with that statement without further description.

The next part – a description of how the policy is translated into local action – is important, because the arrangements can be quite complex – especially in a large organization where the execution of the plan may be significantly devolved. In this case, the policy should describe how the different layers of activity relate, what the plans at each layer comprise (to ensure consistency with the overall plan), how progress to the plans at each layer are reviewed, and who is responsible for the review process at each layer (Figure 3.1).

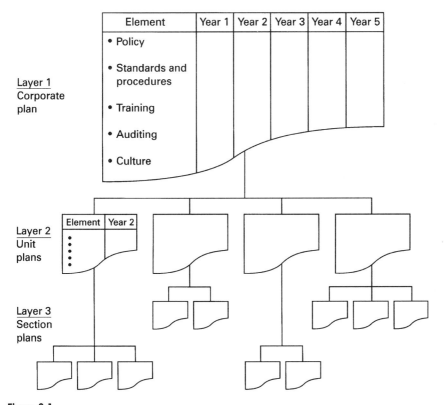

Figure 3.1

Whilst the overall plan may look ahead perhaps three to five years, the subordinate plans need only be annual, reflecting the progress required by the overall plan in that year.

Direct and support responsibilities

Across your organization, there will be a few key people whose contribution will be crucial to the success of your improvement programme. The responsibilities of these people should be spelt out in the policy, making it clear to whom they are accountable. Those with direct responsibilities will be managers with specific remits to ensure that statutory provisions are met – a site manager and site engineer, for example. Those with support responsibilities will be staff in an advisory role – primarily the safety, health and environmental advisers.

Assigning responsibility for acting on the plan at the various layers is crucial to success. Responsibility should lie with the most senior manager of the units at each layer: in the model above this would be the most senior manager in the company for the overall plan, and the Department and Section heads for the subsequent layers. They may delegate the implementation of the plan to suitable subordinates, but should retain responsibility for its delivery, including the periodic reviews of progress, which they should personally lead. Structured effectively, and using pre-determined measures of progress, these reviews need not be lengthy, but the personal leadership and involvement of the managers at the top of each layer is very important. Demonstration of commitment in this way at this level will help to ensure commitment and enthusiasm down the line.

There are just two final elements to this part of the policy: the reviewing and reporting arrangements, and the means of keeping its profile high. The policy should identify the frequency at which progress to the plan is is monitored by those at the head of each layer. In support of these 'high level' review mechanisms, there will usually be progress checks more frequently by those to whom the implementation of the plan has been delegated. These subordinate reviews will vary from one section to another, and do not need to be described in the policy.

Maintaining a high profile for the plan and its progress is a major help in keeping up interest and momentum, so it is appropriate to outline briefly in the policy the arrangements for communicating progress to the plan. Simple routes are:

- through the line management
- through consultative committees
- on display/notice boards.

Consulting and communicating the policy

The policy is the 'launch pad' for your improvement programme. The better your managers and employees understand it, the more they are able to set the more detailed work into context. As we saw earlier in Chapter 2, it is much more difficult to gain commitment to a programme of work if people do not understand how it has been derived. Involving them in its generation is the most powerful process of all. Communication of the policy, therefore, should be carried out as rigorously as possible, using the three mechanisms above.

The primary route should be through the management line, so it is important that managers are briefed in detail about the policy to ensure they can communicate it effectively themselves. Going through this process at the draft stage is the most effective – it allows the opportunity for feedback from staff. This two-way communication process may produce some good ideas which can be used to improve the draft. If the policy has been well written, it is unlikely that there will be significant changes proposed from the communication and consultation process. However, the very act of leaving the policy open for comment and change before final issue will make staff feel a greater sense of ownership to it than would be the case if they had no opportunity to influence.

Once the policy has been established, it is important to keep it at the front of people's thinking. Because a policy document is usually several pages long, it is not practicable to expect that people will be able to retain all the detail, but the policy statement, which expresses the underlying values and principles, should be posted widely as a constant reminder to everyone what those core company values are. The other advantage of this approach is that, although the detail of the policy may change through time, the values statement will be constant.

Auditing and revising the policy

Because the safety policy is the core of your safety improvement strategy, it has to be kept up to date and audited to check that its requirements are working. How impressive is a safety policy which is three years old? What messages does that give your staff and visitors?

Safety policy should be reviewed and reissued every year. There may be very little practical change to the policy year on year, but the very process of reviewing it will make sure that you give some time and thought to whether its component parts are still relevant. Also, it would be unusual if, as you implement your safety improvement plan, you didn't learn something that will help to improve your policy.

Organizational changes inevitably affect your policy – either the management structure which describes the line of responsibility and accountability, or the named responsible people, or both. Annual review of the policy ensures that these changes are picked up formally. Where there are major organizational changes, the policy should be reviewed as soon as possible, even where the annual revision has only recently taken place.

Reviewing and reissuing the policy is rarely a time consuming task, but doing it in a timely and well publicized way gives to your staff a clear signal of your commitment to it. Conversely, leaving the policy to become well out of date sends equally strong signals to those upon whose commitment you rely for the success of your improvement programme.

Periodic auditing of the policy helps to ensure that it is being deployed effectively. It is timely to do this as part of the annual policy revision process, so that any shortcomings may be followed up and, if appropriate, reinforced in the policy. The policy should be audited against best established practice – for example, industry or commercial standards or guides for preparing safety policies – or perhaps, in larger organizations, against an internal standard. The audit should identify, for example:

- the policy's strengths and weaknesses against relevant standards
- whether the named responsible people are in place and are sufficiently qualified
- whether the arrangements for translating the policy into local action are effective at each level in the organization
- whether progress reviews are taking place as described in the policy
- the level of understanding of the policy by staff.

Any significant shortcomings identified during the audit should be followed up by an action plan with clear responsibilities for action, and a means of reviewing progress of those actions to completion.

Policies are the starting point for success, or failure, in safety management. Good policy sets the basis for its deployment. Poor policy is usually an indication of lack of real commitment to managing safety, and seriously constrains any chance of effective strategy and its subsequent management.

Look again at the examples in Chapter 1 if you are in any doubt.

Chapter 4
Assessing hazards and managing risks

'Hazard a guess . . . risk a life'

OK>>>>>>not OK?

In Chapter 1 we explored the difference between small incidents and big incidents and concluded that the underlying causes of big incidents were no different from those of small incidents – the chain of events and the underlying causes can be exactly the same, but the outcomes very different.

Let's take an example. An operator opens up the back of an electrical instrument to make an adjustment, leaving the power on and exposed live terminals inside. The worst outcome of this activity would look something like Figure 4.1.

If Stages 1 to 5 coincide, there is a high probability that the operator will be killed by electrocution. If any of the five events does not occur, the consequences will be less serious. Let us say the operator consistently fails to switch off the power and always uses a screwdriver with faulty insulation, then the chances of serious injury will be the product of the chances of Stages 3 to 5 happening.

If, at Stage 3, the operator's hand slips one time out of five and, when the screwdriver slips (event 4), one time out of four it contacts the live terminal and, when this happens, one time out of ten the operator 'freezes' onto the live terminal (event 5), then the chance of electrocution is:

$$\frac{1}{5} \times \frac{1}{4} \times \frac{1}{10} = \frac{1}{200} = 0.005$$

The chance (it is called the *probability*) of electrocution is 0.005, or one time in 200. Of course, this does not mean that the operator will electrocute himself on the 200th occasion. He may do it on the first (really bad luck!) or the 100th (average bad luck!) or the 300th (relatively good luck!). The other important factor is how frequently the activity is carried out. If the

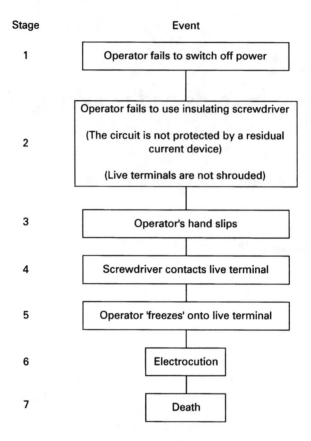

Figure 4.1

frequency is 200 times a month, he's in real trouble! If it is once every two years, the risk of electrocution is much reduced. In the first case the potential frequency is 0.005×200 per month – that is, all things being equal, we will electrocute one operator a month! In the second case, the potential frequency of electrocution is $0.005 \times 1/2$ a year – that is, one operator every 400 years. Clearly, killing an operator a month is not acceptable. But what about one every 400 years. Is that OK?

Let us go back to one of the case studies in Chapter 1 – the *Herald of Free Enterprise*. What were the contributory events? Take a look at Figure 4.2.

Once event 4 has occurred, events 5 to 7 are almost inevitable. How high was the probability of the incident? We don't have any quantitative data. However, we know the ship consistently left its docking position without closing the doors, so we may assume that the probability was 1. We might assume that the steward failing to close the bow doors before the ship left the inner harbour was rare, maybe one in 1000 times – a probability of 0.001. If we assume the supervisor was not there to check that the doors were

Stage   Event

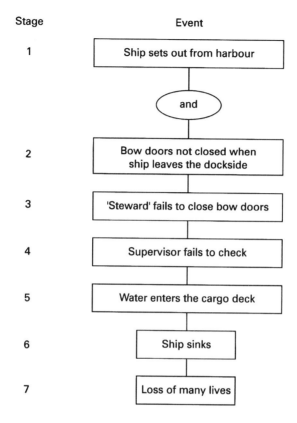

Figure 4.2

closed three times out of ten – a probability of 0.3, then the probability of the doors not being closed at all is 0.001×0.3, or 0.0003 – one in 3300 times. If the ship makes two crossings a day – say, 700 a year, then we might expect each ferry to have a major disaster every 4.7 years (probability 0.0003×700). Good odds for the passengers?

What is the point of calculating the probability of incidents like this? Well, we know that in our operations there are lots of things that can go wrong, but how do we know which of them to deal with first, especially where there may be significant cost involved in taking some remedial action? How do we know we are spending our money improving safety wisely? How do we know which are the big problems and which are the small problems?

When we look carefully at the hazards associated with our operations, it is quite often obvious where the main hazards lie, and what we can do to eliminate or very much reduce them, or to contain them. Where this is simple and cheap to do, then we would just go ahead and do it. This straightforward approach is often known as *hazard studying*. Where your activities are not technically complex, it can be done in a very simple way; where your operations are more technically complex, e.g. running a petrochemical plant, it needs to be done in much more detail.

Unfortunately, life is not always this simple. The consequences of things going wrong, especially where the interactions leading to the consequence are complex, are not always easy to deduce – or, for that matter, how likely they are to occur, and the cost of doing something about those consequences may be very large. In this situation, a more quantitative approach is needed to help decide what action to take, if any. This quantitative approach is often called *risk analysis*, but rejoices under many other names – 'hazard analysis', 'probabilistic risk assessment', 'quantitative risk assessment'.

We will use the terms 'hazard study' and 'risk analysis' to distinguish the two approaches: the first being qualitative, and the second quantitative.

Before getting involved in the detail, let's explore the relationship between hazards, risk, consequences and probability.

Hazards, risk, consequences, probability and frequency

Although we talk about risk fairly freely, we often have problems defining it in terms which help us to understand it. It's not unusual for 'hazard' and 'risk' to become confused – people quite often talk about the one when they mean the other. Let's take a simple view. Here is a hazard.

And here is a 'risk' – you would probably regard it as a 'high risk'!

And here is a 'lower risk'.

A *hazard* is usually defined as 'a chemical or physical condition that has the potential for causing damage to people, property or the environment'. But how do we define *risk*? The dictionary definition of risk is 'chance or probability of danger, loss or injury'. We talk about a 'risky situation' or 'high risk' or 'low risk', but it is almost always a qualitative expression of an unwanted outcome:

high risk = high 'chance' of harm

low risk = low 'chance' of harm

One technical definition of risk is 'a measure of economic loss or human injury in terms of both the accident likelihood and the magnitude of the loss or injury'. However, there are other definitions. One is that risk is 'a combination of incident, probability and consequences'. Confused? Let's go back to basics and take a closer look at the simple sequence of circumstances which gives rise to harm (to property, people or the environment) and what the variables are.

Figure 4.3 is a simple portrayal of a combination of a hazard and an incident to produce an unwanted consequence. We'll explore each of the factors in turn, and the variables associated with them.

Figure 4.3

Hazards

In any given situation, the hazards are fixed. They may be such things as flammable liquids, toxic substances, potential energy. They can vary in two

Hazard	Incident	Consequences
Flammable substance	Ignition	Fire
Toxic substance	Human exposure	Health effect
Corrosive substance	Uncontrolled release	Environmental damage
Road travel	Collision	Injury/vehicle damage
Deep water	Submersion	Drowning

Figure 4.4

ways – their intrinsic *nature* (e.g. high flammability/low flammability, high toxicity/low toxicity, high pressure/low pressure) and their *scale* (a lot or a little). Take, for example, a flammable liquid. At the top end of the hazard scale, we might have a 1000 tonne storage tank of highly flammable liquid, and on the lower end a 500 ml bottle of low flammability liquid. Hazards are *potential* problems.

Incidents

An incident is an event which realizes the damage potential of a hazard. Obviously, an incident has to be relevant to the hazard: in Figure 4.4 there are some examples, with the consequences.

Incidents can be quite simple, but more often they are complex, with a number of events combining to cause the incident. Remember the major accidents described in Chapter 1? In all these cases there were many events underlying the eventual incident. On Piper Alpha, for example, the incident was the ignition of flammable gas, but there were many contributing, sequential events. Figure 4.5 is a partial analysis of the sequence and combination of events that led to the escape, and ignition, of flammable condensate on Piper Alpha.

You can see from this 'event tree' that removing any one of the events could prevent the incident. For some hazard types, there are different types of incidents which would lead to different consequences. For example, an escape of corrosive substance could, under different circumstances lead to:

- human exposure
- environmental exposure
- equipment exposure

and we could construct event trees which would describe events leading to any of these three.

Incidents can not only vary in *type* but in *frequency*, and in risk analysis the frequencies and probabilities of events leading to an incident are assessed to determine risk. More on this later.

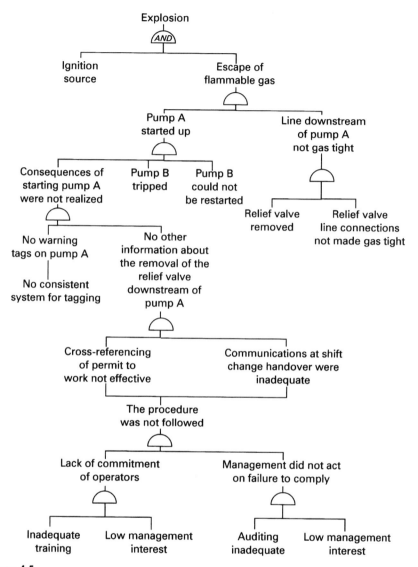

Figure 4.5

Consequences

Consequences can vary in their severity (how bad) and frequency (how often), and it is the combination of these two factors that dictate the way in which we view the acceptability or unacceptability of the risks.

Going back to the flammable liquids case, the *consequence* of the loss of containment of flammable liquid and an ignition source is fire, and the consequence of fire is property damage or injury or both. Whether the consequence is major or minor is dictated by two things. The first is the *scale* and *nature* of the hazard: a fire involving a lot of highly flammable liquid, for example, will have a much bigger consequence than a fire involving a small amount of low flammability liquid. The second is the *type* of incident – small loss of containment versus major loss of containment. Of course, if we also calculated that the probability of the consequences being realized was every month, say, we'd be a lot more concerned than if the probability was once in 1000 years.

Just to complicate the picture even further, there can be primary consequences and secondary consequences. If you recall the Space Shuttle *Challenger* incident, the consequence of O-ring failure was escape of hot gases from solid fuel combustion (primary consequence), followed by the failure of the hydrogen tank and escape of hydrogen (secondary consequence) which inevitably led to the explosion which destroyed the shuttle.

Clearly, it is important to understand *all* the potential consequencies of hazard/incident combinations. Earlier (Figure 4.5) we looked at a method of identifying the events that would lead to a particular incident – an 'event tree' or 'fault tree'. This structured approach may be used to identify both primary and secondary consequences of hazard/incident combinations, especially where you want to identify the 'worst case' consequence.

There is obviously a wide variety of consequences dependent upon the nature of hazards and events. Some of those hazard/event combinations will lead to small consequences, some of them to major consequences. Some of these combinations may be fairly frequent, some very rare, some in between.

Probability and frequency

Probability and frequency are the two terms used in calculating risk. The two are different, but are commonly used together. Probability is the chance of an event occurring, measured on a scale of 0 to 1, where 0 represents no chance of occurrence and 1 represents absolute certainty of occurrence. Probabilities are often used as an expression of the reliability of hardware, e.g. if an electronic instrument is known from test data to fail one time in 100 000 demands, then its probability of failure is 10^{-5}.

Frequency is simply a rate, e.g. 25 times a year, and is calculated from best known data. This may be from historical data, from test models, or from extrapolation of historical incident data to new situations. For these reasons, there is often a degree of uncertainty about event frequencies, particularly when you take into account equipment ageing and the imprecise nature of, for example, human error data.

Cast your mind back to incidents and events. If every event leading to an incident has known frequency or probability, the incident frequency may be

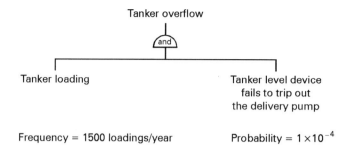

Figure 4.6

calculated. For example, in the very simplified event tree in Figure 4.6, the frequency of tanker overflow is 1500×10^{-4} spillages / year, i.e. one spillage every 6.7 years. In this case the magnitude of the overflow will depend on other circumstances, e.g. human intervention to trip the delivery pump.

Back to *risk*!

We have explored the simple model in Figure 4.3 and its underlying complexities, and I hope it has become clearer that where we need a *quantitative* view of hazard/incident/consequences combinations, a knowledge of two things is essential: how serious (are the consequences) and how often (will it happen).

'How serious' can be measured in many ways, e.g.:

- number of employee fatalities
- cost of damage
- number of serious ill health cases in the local population.

There are other 'how serious' factors which may be less easy to quantify, but which are still highly relevant to the continuation of your operations – the response of enforcing authorities and local communities, for example.

'How often' is simply frequency (number of times a year, for example). So we can measure risk in terms of the combination of the two. For example:

- number of employee fatalities per year
- financial loss per year.

Building on Figure 4.3, therefore:

hazard + incident = consequence . . . and . . .

consequence × frequency = risk.

Because there are so many combinations of hazards, incidents and consequences, it is impossible to assess them all; in any event, we would want to avoid taking a complicated quantitative approach when a simple qualitative approach will do. So let's explore a simple guide to hazard study and risk analysis which helps to apply a 'simplest effective solution' approach.

The guide comprises the following

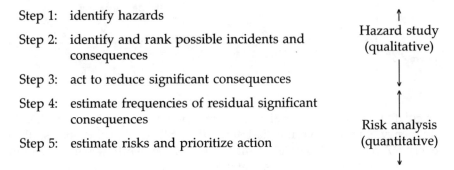

Step 1: identify hazards

Step 2: identify and rank possible incidents and consequences

Step 3: act to reduce significant consequences

Step 4: estimate frequencies of residual significant consequences

Step 5: estimate risks and prioritize action

↑
Hazard study
(qualitative)

↓

↑
Risk analysis
(quantitative)

↓

Step 1 Identify hazards

Hazard identification is relatively straightforward, and can be aided by using a check list of the type in Table 4.1.

Table 4.1 Checklist for hazards

Chemical and biological	• flammable
	• explosive
	• unstable
	• highly reactive
	• toxic:acute/chronic
	• corrosive
	• irritant/sensitizer
	• biologically active
	• asphyxiant
	• environmentally harmful
	• pyrophoric
Physical	• kinetic energy
	• potential energy
	• radiation (ionizing/non-ionizing)
	• noise/vibration
	• physical stress
	• heat stress
	• electrical
	• vacuum
	• high/low temperature

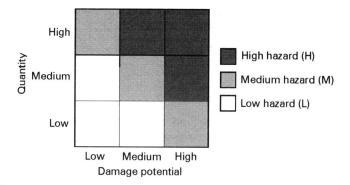

Figure 4.7

Listing of hazards against your checklist should identify important features such as

- quantity
- physical state (dense solid, particulate solid, liquid, gas) and physical characteristics (melting point, boiling point, flash point, smell threshold, etc.)
- location.

Where you have chemicals, it is important to identify any combinations which could lead to a hazardous situation, e.g. peroxides and organics. At this stage you may wish to identify broadly the main areas of concern, characterized by Figure 4.7, where the damage potential may be expressed in terms of flammability (measured in terms of flash point), toxicity (measured in terms of exposure limits), potential energy (measured in terms of pressure), and so on. The main emphasis will be on those combinations of high damage potential and quantity, which fall into the 'high hazard' area.

Step 2 Identify and rank possible incidents and consequences

Working from your knowledge of the hazards, you are able to identify possible incidents and consequences – and, extending the picture, the hazards and events giving rise to the incidents. If your operating hazards are relatively simple, you should be able to identify associated incidents and consequences from your operational knowledge and, where necessary, construct simple event trees from any of the significant consequences to start to identify the key events that may be influenced to reduce those significant consequences. A rough illustration of this approach is in Figure 4.8.

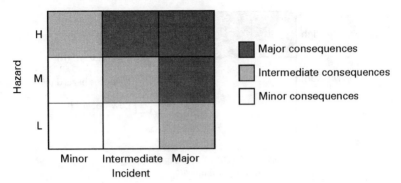

Figure 4.8

The hazard scale of *High*, *Medium* and *Low* is taken from Figure 4.7 and the scale of the incident is judged by its severity, e.g. a loss of containment incident may vary from a small leak on a pipe flange (minor) to fracture of a 100 mm transfer pipe (intermediate) to failure of a storage tank (major). Using a simple ranking arrangement of this type allows you to concentrate on the hazard/incident combinations leading to major consequences.

Where your operations are rather more complex, you may need to adopt a rather more structured approach which helps examine each potential consequence:

- fire
- explosion
- human exposure to chemical/biological agents
- environmental damage
- human exposure to physical agents
- energy release.

With a knowledge of hazards (Table 4.1) and their locations, for each potential consequence in turn you ask:

- how can it occur?
- where can it occur?
- what are the primary and secondary consequences?

Where you decide the consequences are significant, you should act to eliminate or reduce them where this is cost effective (see step 3). Where this cannot be achieved cost effectively, you should move into a more quantitative assessment to assess the frequency of the consequences (step 4).

A more searching technique still is a 'hazard and operability study', used typically for chemical and petrochemical plants. This technique is applied to the engineering line diagram of the process, where process flows are known and operating and maintenance procedures have been outlined. The engineering diagram is broken up into manageable parts, and each part examined in turn. The study team works to a check list which explores deviation from the design operating intent of the plant (flow, temperature, pressure, etc.) asking how the deviation can occur and what the consequences of that deviation would be. Significant consequences are addressed in the way described earlier. This technique has the advantage of systematically identifying not only those events which have safety, health and environmental implications, but those which have implications on the effective and simple operation of the process. It is, however, a rigorous procedure which requires expertise and experience to be applied effectively.

Step 3 Act to reduce significant consequences

Going back to our simple model:

Hazard + Incident = Consequence

we can eliminate or reduce consequences by acting on each of the three elements:

(a) eliminate/reduce hazards
(b) eliminate/reduce incidents
(c) mitigate consequences.

They should be tackled in that order, since dealing with a problem at source is always more reliable as a solution, and usually the most cost effective option.

Eliminate/reduce hazards

Questions to ask here are:

● can I eliminate the hazard?
● can I substitute it with something less hazardous?
● can I use less?
● can I do it more efficiently?

The ability to apply these simple questions effectively usually depends on the complexity and nature of your operations. If you are operating a solvent degreasing plant, maybe it would be easy to move to a less hazardous solvent. If you are operating a chemical plant, however, the situation is likely to be more complex, and there may well be a trade-off between the process chemicals, process conditions, operating complexity and process waste. Sometimes, it helps to turn conventional thinking on its head and ask how it could be done better – by using different technology or by switching to a different product which involves less chemical and process hazards and which has a wider market application; however this will usually entail shifts in strategic thinking and will have to be balanced against potentially significant capital cost.

Sometimes, simple things can be done to reduce hazards – splitting down bulk hazardous storage into smaller units, or reducing total storage by having more frequent deliveries, for example. Quite often we don't see opportunities until we challenge the *status quo*.

Eliminate/reduce incidents

When you have constructed an 'event tree' of the incident, you have a list of the events which contribute to the incident, allowing you to explore ways of eliminating or reducing the likelihood of the events occurring. Options here could be:

- use of more reliable equipment (less probability of failure)
- better maintenance of equipment (less probability of failure)
- simplification of processes (less opportunity for failure)
- reduce human error (less opportunity for failure)
- physical separation of events whose combination is key to the incident (e.g. separation of ignition sources and flammable materials; physical interlocking of critical valves or switches).

Sometimes the processes we operate are more complex than they need to be, often resulting in more opportunities for things to go wrong. Job Safety Analysis is a very powerful methodology for identifying opportunities for simplifying operations, resulting in less opportunity for a hazardous event. This leads us into the frequently overlooked area of human error, which is often a significant contributor to incidents. Here, the issues are very basic:

- how complex is the operation?
- how frequently does the operator do it?
- how well has the operator been trained?
- how motivated is the operator to follow the correct procedure?
- how well does the operator understand the consequences of getting it wrong?

In most organizations, there are big improvements to be made in this area, and they are explored in other parts of this book.

Mitigate the consequences

After having done all you can to eliminate or reduce hazards and incidents, the next stage is to explore ways of mitigating the consequences. This almost always involves expense – another good reason for as intrinsically safe a design as possible. Two examples are building containment walls (for example to contain explosion or fire or spillage) and installing automatic fire extinguishing systems.

Step 4 Estimate frequencies of residual significant consequences

When you have done all you can to reduce the consequences of hazard/ incident interactions, and you are still left with unacceptable consequences, you need to understand what risks those consequences present – that is, how frequently will they occur. This is where risk analysis techniques apply. These techniques are more detailed applications of the basic 'event tree' or 'fault tree' analysis approach described earlier in this chapter, using frequency, probability and reliability data to calculate the frequency of events, incidents and consequences. I shall not attempt to describe the techniques in any detail here: there are comprehensive texts on risk analysis, and the application of the methodologies require expertise and experience – it is not a DIY job!

Step 5 Establish risks and prioritize actions

Risk analysis quantifies the risks associated with significant consequences, helping to prioritize action. The outcome of risk analysis could typically produce a ranking of risk in the form described in Figure 4.9, which will vary dependent on the scales used.

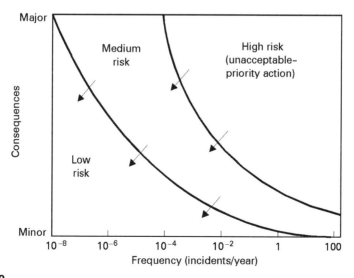

Figure 4.9

The range of consequences can be described, for example, in terms of injuries/deaths, property damage or environmental damage. Priority action would be given to high risk situations to bring them down into the 'medium risk' band, and subsequent efforts would be targeted at moving medium risks down towards the low risk band. In doing this, cost benefit analysis should be applied to make balanced judgements for improvement based on cost and associated reduction in risk.

Conclusions

This chapter is designed to give the reader a reasonable understanding of the key features of practical hazard studying and risk assessment. It is a major area which uses detailed and sometimes complex methodology, which can only effectively be applied by experts. However, it is worth bearing two things in mind. First, there is a lot that non-experts can do using simple, structured analytical techniques to identify hazards and incidents and unacceptable consequences, allowing good judgements to be made about the reduction of risk without needing recourse to detailed quantitative techniques. Second, risk analysis is not a precise science, and its conclusions are not absolute; it is, however, the best we can do with current experience and it is valuable in estimating levels of risk and in providing comparisons of risk to help focus effort on those risks which are most significant.

I make the point again that the human contribution to the 'incident' part of the hazard–incident–consequence sequence is often very significant and very difficult to quantify. This makes the often neglected area of human behaviour such an important issue.

Finally, we should remember to learn from experience. We may be stuck with the processes we've got, and we do our best by all the 'retrofitting' described in this chapter. There is no excuse for making the same mistakes when we set up our next operation, or build our next plant, however, and it makes good business sense to make them as *intrinsically* safe and environmentally friendly as we can.

Chapter 5
Setting standards

'Show me your standard, and I'll show you mine . . .'

Standards – who needs them?

Many people have a natural aversion to the concept of standards. It conjures up pictures of bureacracy, conformity, lack of individual creativity and expression. So who wants standards? There are national standards (British Standards in the UK), European standards and international standards for all sorts of products and processes. We have had British Standards for many years. In the UK, one of the first British Standards reduced the number of different tramway rails from 75 to 5, saving a million pounds a year even then. During the First World War, standards helped to make sure that fighter planes were manufactured economically and quickly, and that shells fitted gun barrels!

In our fiercely competitive commercial environment, we need to be assured that products that we use are manufactured to a minimum specification, either to ensure that they will be safe in use, or that they will work reliably in use. If, for example, you were buying an electrical plug and you had the choice of buying one made to the British Standard and one which was not, you would prefer the former, even though you may pay more. The reason? Because buying the British Standard plug would give you assurance that it would be safe to use; you would have no such assurance with the other. In fact, if it wasn't British Standard kitemarked, you might be sure that it didn't meet important levels of safe design and manufacture.

Without standards for the design and manufacture of products, we – the consumer public – would not be able to choose with confidence which of any number of competitive products we should choose to buy to guarantee that the one we choose will work – and work without putting us at unnecessary risk.

The important role of standards is that they capture the best essential features for safe and reliable use. So how does this apply to the safety of our operations? Well, reflect on why we write down ways of doing things. It is because we have learned from experience that some ways will work well and some not so well. In a safety context, those ways that do not work as well have probably resulted in injury or damage, so it is not in our interests to repeat those failures. From both an operational safety and commercial point of view, it *is* in our interests to follow the ways that we know work well. In the various operating parts of organizations, these 'ways that work well' are usually known as operating procedures. Some operations are common right across the organization, but it is not unusual to find that the operating procedures describing these common operations are different from one operating unit to another. Does that matter? Maybe, maybe not. It depends on whether the difference is significant in terms of best known practice. But how do you decide what *is* best practice? As far as each operating unit is concerned, *theirs* is best known practice.

In these situations it is common to find that none of them is best practice. They will all have some good features, based on their individual experience – but those experiences are not collective. This is where corporate safety standards have their place – to act as the point of collective experience – as a source of best known current practice.

Standards are different from procedures. A standard describes *what* has to be achieved, and they key features of known best practice. A procedure is the local application of the standard, and describes *how* it is achieved. That combination of corporate best practice and local circumstances is a very powerful one for local procedures. Another advantage of corporate standards is that they take a lot of the work out of writing local procedures since, apart from local needs, all the necessary information is contained in the standard and its supporting best known practice.

A further important role of safety standards is to act as the 'corporate memory'. When accidents happen, it is sometimes found that the established controls are inadequate – for example, a design feature, materials of construction, operational limits, operational controls. The unit in which the accident occurred will learn from it, and may alter local procedures to avoid a recurrence. But how is that learning captured in a way that any similar incident is avoided *anywhere* in the organization, and for *always*? Too often when accidents occur in organizations, someone will look back and see the same accident having occurred before – maybe more than once. Reflect back to the major incidents in Chapter 1. In all these cases, the systems failures which led ultimately to each disaster were repeats of what had gone before.

So, safety standards (and their supporting best practice) are important where variation could result in significant injury or damage. They provide for a 'corporate memory' of the means of avoiding those incidents which have gone before, and they enable a consistency of safe systems of operation throughout the business, based on local procedures.

Writing the standards

Providing company safety standards is no small exercise, and should be led at senior management level with reviews of progress at the highest level in the organization. The exercise must be owned by management with strong support by professional safety managers and specialists.

When do we need procedures?

One of the reasons why we have a natural aversion to procedures is that we come across so many. Sometimes we encounter dossiers full of procedures, some of which are important, some less so. Some local procedures are based on those company standards which are relevant to local operations, and some based on specific local needs. In these situations, the less important procedures are taking our eye from the more important ones. There may, of course, be a good case for all the procedures, but why give them all equal weighting? If I gave you my safety procedures manual containing 100 procedures, how many would you read before losing interest? However, if I picked out the 10 most important, you would be more likely to read them. How do we discriminate?

Consider the matrix below (Figure 5.1). If we consider the whole spectrum of hazards in our operations, and the likelihood and consequences of their occurring, we would seek to control the different levels of risk in different ways. Very high-risk circumstances (combinations of

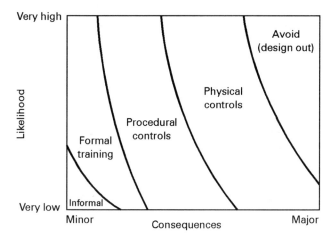

Figure 5.1

high likelihood of occurrence and high consequence) are normally avoided – for example by designing them out of a process. Slightly lower, but still significant, risks would normally be subject to physical controls, for example interlocks on machinery guards. Some risks do not warrant physical controls, but still require some form of control because variation in the way in which they are controlled could lead to serious injury or damage – this is where formal procedures apply. Then there are those low risk circumstances where only very simple controls are needed – for example the use of a knife in catering operations: this is where on-the-job training applies.

So, the extent of control should be dependent on the risk. Think back to the incidents in Chapter 1. Take the *Herald of Free Enterprise*. What type of control would be appropriate to ensure the bow doors were closed before leaving harbour? Are procedures enough? Clearly not. In this case physical controls would have been appropriate, such that, for example, power could not be transmitted to the ship's propellers whilst the bow doors were open or, at the very least, some sort of fail-safe warning signal on the bridge to indicate when the doors were open. In the light of more recent incidents, it is being questioned whether the design of roll-on, roll-off ferries is intrinsically safe, and whether the risk of water ingress into the cargo decks should be designed out altogether – the most rigorous form of control in our matrix.

The same question might be asked of the space shuttle *Challenger* incident. Was the seal between the booster rocket motor sections a sufficient physical control, or was there an intrinsic weakness in the arrangements which should have been designed out ('avoided' in the matrix)?

The immediate causes of some of the other major disasters were clearly failures to follow established procedures. Procedures were certainly an appropriate form of control to carry out the signalling rewiring work (Clapham Junction). But if the consequences of a single failure in all this extensive rewiring work (and we could reasonably argue that the likelihood was reasonably high) could be seen to result in a signalling failure (again, a fairly high likelihood) which in turn could result in the failure of a train to stop (consequences 'high'), we might question whether additional physical controls would have been appropriate.

Many processes require human controls – it simply isn't possible to do everything automatically. In these cases we need to ask ourselves what are the consequences of failure: what is the worst thing that could happen, and what is the likelihood – as best we can predict it? If, on this basis, we judge that human controls (through procedures) simply are not reliable enough, we have to establish a second line of defence – either physical control or, if this simply is not possible, an additional layer of *highly reliable* human control (for example, a good deal more reliable than the supervisory rewiring checks at Clapham Junction!).

What are the features of good procedures?

If you know someone who likes writing procedures, you have a rare find – it is not a job for which there are many volunteers! For this reason, it is not unusual to find that procedures gradually grow out of date, and it becomes a marathon exercise to rewrite them. How do we:

● make them as easy as possible to write in the first place?
● co-ordinate the whole exercise?
● keep them up to date?

Where you are faced with a big task of producing safety procedures, there are ways of minimizing the effort involved, but no way of actually avoiding it. So, sharing the task as widely as possible can have real attractions. In principle, this approach should speed up the production of your procedures, but in practice it doesn't always work out this way. Why not? The two main reasons are commitment and personal style. If you have not got the commitment, no amount of cajoling is going to get the procedure written quickly. If each procedure is drafted in unique personal style (and we all have our own individual approach), it can lead to a major rewriting job which defeats the object of sharing the work around. So, is it better to have a hard core of procedure writers, or is it better to share the work around, and how do we overcome the 'commitment' and 'personal style' problems?

Having a hard core of writers will certainly ensure a consistent style, but in order to achieve the job in a reasonable period of time, the writers will have to be given the time. A lot here depends on the level of priority that is given to this task; it simply isn't possible to fit in this job with all the others. Specific blocks of time have to be put aside to achieve real progress. Remember, too, that 'procedure fatigue' can set in – people can only take so much procedure writing at once! Where procedures are very much out of date, you can almost guarantee that rewriting them raises questions of procedure that have to be resolved, and this can take up considerable time.

Another option is to contract out the procedure writing. Again consistency of style should be good, but there is a risk that the procedures will lack local 'flavour' unless the writers take the trouble to understand the local culture – and this can take some patience and time.

If you opt to share the job around your organization, what are the benefits, and how do you overcome the problems mentioned earlier? The first advantage of this approach is that you can bring local knowledge and expertise to bear on each procedure. This is particularly relevant for local procedures which are not subject to company standards. The second advantage is that if you choose your authors from amongst those who will have to implement the procedures, it serves as much as a personal awareness exercise as a procedure-writing exercise. Of course, people who write procedures generally feel a greater sense of ownership and commitment to them than they would if the procedure had been written by someone with less local knowledge. A really powerful combination is when procedure drafting is subject to local discussion as this then doubles as a local training exercise. Earlier, I raised the question of the commitment of people given the task of writing procedures. The only way around this problem is to give it a really high profile, letting the authors know that it is an important task, to be completed to deadline, and letting their managers know that they need the time and encouragement to carry out the task to schedule. Extra impetus may be gained by giving the whole exercise a high profile – for example by measuring and publicizing progress.

Achieving reasonable consistency in style, especially when the task is shared amongst a lot of people, is a real challenge. The first thing that can help here is the availability of company standards; they make it so much easier to write the corresponding local procedures. The second thing that can help is a very clear guide as to how the procedure should be written. Here are some key pointers.

- say why it is necessary
- concise (minimum words)
- simple language (no jargon or abbreviations)
- clear distinction betwen the mandatory and the advisory
- use pictures (flowcharts, for example) rather than text, wherever possible
- make responsibilities clear
- identify training requirements
- make sure it is auditable: include a basic audit check list, if possible.

There are different options to the way in which your local procedures are written – you have to decide which is the most appropriate for you. When considering your options, remember to take into account:

- the need for local knowledge
- how quickly each option can deliver
- the advantages of local involvement
- cost (hiring expertise, overtime, etc.).

Co-ordination

Writing for the first time, or rewriting, procedures is a major exercise usually needing close co-ordination. An effective approach, which assures management ownership, is to appoint a suitable senior manager to lead the project and to have the freedom to appoint a project team which will act as an editorial board. The team will be responsible for setting the style in which the procedures should be written, for making a considered choice of authors, for briefing authors in such a way that they clearly understand the required content, style and deadline, for editing the draft procedures, and for monitoring progress to schedule. The project team leader should be responsible for reporting progress to the local head of operations. Anything less than tight co-ordination will almost inevitably result in programme slippage.

Keeping them up to date

Now that you have completed the major task of producing your local safety procedures, how do you avoid having to go through the same exercise again in five years' time? There are two things that can help. The first is by having a management-led review team, responsible for keeping the procedures under review year by year, and by having this team report back to the head of operations as a specific part of the safety improvement plan. This way the procedures continue to receive specific attention, and are much less likely to fall into disrepair. This review mechanism should also take into account any incidents that have occurred which call into question the completeness of the procedures and, where relevant, its corresponding company standard.

The second is through safety auditing, which we will look at in Chapter 7. Auditing helps not only to identify how effectively local procedures are being practised, but whether those procedures continue to be reasonable and workable under local circumstances which may change through time.

Procedures, practice and common sense

Accident case histories are littered with examples of people just not following procedures. Let's take a look at examples from the major incidents described in Chapter 1: the quotes are taken from the judicial enquiry reports.

Clapham Junction

This incident is rich in failures to take safe systems of work seriously – from the delay in issuing important procedures, to the failure to communicate procedures, to the failure to see that they were in any way working. We can start with the basic wiring design.

> 'The procedure for specifying rewiring details was that once the drawings were ready, the Design Office would issue three copies

to the Signal Works Assistant. He retained an office copy for later use by the Testing and Commissioning Engineer and passed a second copy on to the supervisor and the third to the senior technician. The evidence showed that this proper procedure was frequently not followed.'

' If any problems arose on the ground in translating the intentions of the design office as shown on the wiring diagram into the reality of the wiring in the relay room, then a practice of informal consultation between the design office and the. . .technician involved came into operation.'

This informal consultation was a feature of the rewiring job which led to the signalling failure.

'The informality and imprecision of these arrangements meant that the arrangements themselves had within them the potential for confusion and misunderstanding. They also had the potential for creating future problems if, as in this case, the wiring diagram (which is kept in the relay room after the work had been done) did not accurately record the wiring which had actually been carried out on installation.'

Modifications to procedures should be treated with the same seriousness as the original procedure itself. In this case the informal modifications could well have caused later confusion to anyone carrying out wiring checks.

There was also an important procedure requiring that an independent wire count be carried out to confirm that what was physically in place after a new wiring installation corresponded to the wiring diagram for the whole installation.

'He (the engineer carrying out the rewiring job) said that he did not think he was really being supervised that day, because (his supervisor) was working outside. He was not expecting anyone to check his work visually that day. He was not expecting anyone to do a wire count, that is to say to do an independent check whether the right number of wires according to the wiring diagram were physically in place at each point in the installation. In his experience on the changeovers he did, nobody ever did a visual check of his work. Nobody ever counted the wires after him.'

The management structure failed to carry out what was known to be standard good signalling practice. The extent of this failure raises some fundamental questions about many of the key elements of any safety management arrangements. Were the roles and responsibilities of managers clearly defined? Were adequate resourses provided to enable the requirements of the procedures to operate effectively – in this case the wire counts?

Were the communication and training arrangements good enough to ensure that people knew what the procedures were and the importance of carrying them out diligently? Were there any mechanisms to check whether procedures were being followed (safety auditing)? The answer to all these questions is that there were significant deficiencies in them all.

The following extract from the investigation report about the way in which formal procedures can be overtaken by custom and practice makes sobering reading.

'(The technician) was considered by his supervisors to be a very good worker. He was the sort of person who could be left alone to get on with the job in his own way. He would make his own assessment of the task in hand. He would consider it carefully. He would make a plan in his head, and then put it into effect. There were no complaints about the standard of his work. At all times, over the years, months and weeks leading up to the accident, the general picture of (the technician) in the eyes of his colleagues and superiors was that of a thoroughly competent and efficient senior technician.

The reality, sadly, was very different from the picture. Many of the errors (the technician) made in the relay room on Sunday, 27 November 1988 he had been making all his working life. They were not isolated, momentary lapses, they had become his standard working practices.'

These practices resulted in disconnected wires not being cut back (to prevent them from remaking contact), not being tied back, and not having their bare ends covered with new insulating tape – all requirements of rewiring procedure. The report points out that this technician's failures to follow procedures was not an isolated case:

'Still more disturbing was the fact that these errors of practice were not (the technician's) alone, but were. . .part of a wide-spread way of working, almost a school of thought, at technician, senior technician and even supervisor level. . .'

With this sort of culture in place, the technician's failure to follow established procedures becomes much less surprising. This general replacement of 'company standards' with custom and practice was reflected, as the report points out above, at supervisor level.

'(The technician's supervisor's) reasons for not doing an independent wire count in the relay room were: (i) that he was expecting (the technician) to wire count his own work as he went along; (ii) that the general practice for years back had been one of installers checking their own work; and (iii) independent wire counts had not been mentioned by anyone as a requirement for (rewiring) work and he had never known an independent count to be done in relay rooms on commissioning days.'

This lack of clarity and commitment reflected itself in higher levels of management. The testing and commissioning engineer responsible for carrying out final functional tests and certifying installations ready for service did not understand that these requirements related to him. He believed that the departmental instruction SL-53 on the 'testing of new and altered signalling', which stated that 'a wire count must be carried out on all free-wired safety relays and terminations and recorded on the contact/terminal analysis sheets' was still a provisional requirement and that, in any event, he regarded wire counting to be the responsibility of supervisors and leading technicians. In reality, *nobody* thought it was their responsibility to carry out an independent wire count, and *nobody* from the testing and commissioning engineer downwards had a concept of the importance of an *independent* wire count.

The instruction SL-53 had previously been issued in draft form, but had been given full authority as a departmental instruction three years prior to the incident. The letter giving this authority, however, had not been received by the testing and commissioning engineer's predecessor, who did not implement the instruction. This failure in communication was not an isolated incident. As the report points out:

'How it could come about that a tester on such safety-critical new installations. . .could be happy to turn his back on so fundamentally relevant and important an instruction as SL-53 would seem inexplicable were it not for the realisation of how poor were BR's channels of communication and instruction in the S&T department of southern region.'

Another extract from the report says something about the overall looseness of attitude to important procedures:

'Despite the fact that (the testing and commissioning engineer's predecessor) was appointed as testing and commissioning engineer for the whole of the south western area on reorganisation in May 1988 and served in that post until 3 October 1988, he had never even read SL-53. Although he had had a copy from about mid-1987, a year before his appointment, and although throughout that year he had been responsible for all the WARS (Waterloo Area Resignalling Scheme) testing, amazingly he had only ever glanced through SL-53. He did not like it and he did not attempt to work to it.'

In retrospect, it seems incredible that there were so many failures to apply such an important safe system of work, but it also reminds us how easy it is – especially in a large organization undergoing change – for basics such as this to be neglected if effective safety management systems are simply not in place.

Piper Alpha

Operating an oil platform in the North Sea presents special problems. The first is the very nature of the operations themselves – extracting large quantities of highly flammable gas and flammable oil from below the sea bed in relatively hostile conditions gives special emphasis to the need for safe systems of work. The second is the limited means of escape, which gives special importance to the emergency procedures. Because of the importance of maintaining the integrity of engineering arrangements used to control the withdrawal of oil and gas, a 'permit to work' procedure was used to control any engineering modifications or maintenance work. So, in view of its crucial importance, how well was the permit to work procedure complied with?

To understand the failures of the permit to work system on Piper Alpha, it is necessary to have a broad understanding of how the system was supposed to work. It is designed to ensure that work which, for safety reasons, needs to be controlled is planned, with precautions, protection needs and responsibilities defined and written down and communicated to those who are part of controlling the work. Some of the more specific requirements were that:

'– the equipment to be worked on is identified, the maintenance work to be done is specified and approved by a senior supervisor, the Approval Authority
– the equipment is isolated from the rest of the process and remains so for the duration of the work and the safety precautions necessary for the work, such as gas testing or use of protective clothing, are listed prior to the issue of the permit; these actions are the responsibility of the Designated Authority
– the maintenance work is carried out as specified by the permit, the safety precautions as listed are adhered to, and upon satisfactory completion of the work the permit is returned to the Designated Authority; these actions are the responsibility of the Performing Authority
– finally, the equipment is checked to confirm that the work has been satisfactorily completed and the isolations removed so that the equipment can be returned to service; these actions are the responsibility of the Designated Authority.'

Occidental operated a PTW (permit to work) system on Piper Alpha in which the Approval Authority was the production superintendent, the Designated Authority the shift lead production operator, and the

Performing Authority the shift maintenance lead hand or, alternatively, the supervisor of a group of contractors' personnel.

When maintenance work stopped for any length of time – overnight, for example – the PTW was supposed to be suspended by the Performing Authority returning the permit to the Designated Authority and both signing that the work was suspended. During this suspension the equipment would remain isolated, and the permit would be reissued when work was to be resumed. Information on maintenance work was supposed to be included in the handovers which took place at shift changeover, and in various logs.

The report of the enquiry identified that '. . . in a number of significant respects this (the permit to work) procedure was habitually or frequently departed from. . .' and gave some specific, important examples.

- The procedure required that the Performing Authority took the permit to the Approving Authority in person. This was often not done.
- The section of the permit which asked whether there was any other work which might affect the work covered by the permit was seldom used. At best it was ticked, but with no supporting detail.
- Multiple jobs were undertaken on a single permit, contrary to the requirements of the procedure.
- When Performing Authorities returned permits to the control room shortly before the end of the day shift, they would sign off all copies of the permit and leave them on the desk of the lead production operator. This was again contrary to the procedure, which required the Performing Authority and Designated Authority to meet. This practice had developed because the lead production operators were out doing their handovers. It was also the case that Performing Authorities did not always check the work to ensure it was in a safe condition to leave overnight before handing in the permit for suspension, again contrary to the procedure.
- Suspended permits were filed in the safety office overnight, although the procedure required the Designated Authorities to retain them. As a result, a night shift lead production operator would not know which permits had been suspended, and what equipment had been isolated for maintenance.

These are just some of the examples of the many failures to follow the PTW procedure which, as a result, could not provide that degree of defence which was vital to the safety of maintenance operations.

There were also serious questions about the effectiveness of the PTW procedure itself, and whether it met best industry standards at the time. Some of the points made by the report were as follows.

- Apart from major shutdowns, there was no consistently used system for tagging isolation valves which had been closed as part of the arrangements for isolation of equipment for maintenance, with the objective of warning that the valves should not be opened during the maintenance work. There was also no consistent practice for physically locking off isolation valves which had been closed for maintenance work, and even where equipment had been locked off, there was nothing to tell the operator what the reason was.

- Where the work under one permit could affect the work under another, there was no cross-referencing of the two.
- The PTW suspension arrangement led to large numbers of suspended permits, some of which had been suspended for months, and the correlation of these permits with active permits was difficult because they were filed not according to location but to the trade involved. Needless to say, it was not easy to see which equipment was isolated for maintenance.

The report drew attention to other questions of adequacy, but those above are enough to make the point that there were weaknesses in the procedure itself, even before the question of adherence to the procedure arises.

There were inadequacies also in the other key procedural arrangements which should have protected platform operators from the consequences of an incident on the platform – the emergency procedures. The first inadequacy was that the platforms Claymore and Tartan, which fed gas to Piper Alpha, were not prepared for an emergency on another platform. Because of this, even after they received the mayday call from Piper Alpha, they continued production in the belief that Piper Alpha would bring the incident under control. As a result, the compressors on Tartan exporting gas to Piper Alpha were not shut down until 15 minutes after the initial explosion, just five minutes before the second major explosion caused by the rupture of the gas riser from Tartan to Piper Alpha.

Another key procedure – emergency evacuation – was not adequately understood, although this relates more to failures in training, which is explored in detail in Chapter 6.

What are the lessons from these and many other incidents where procedures designed to assure safety of operation so signally fail to provide that protection? The first is that procedures should be consistent with best known practice. It simply doesn't make sense to disregard good practice that has been adopted by industry and commerce, especially when it has often emerged out of the learning from significant incidents.

The second is that no matter how good a procedure is, if it is not communicated well and reinforced by management, we cannot reasonably expect people to understand it and take it seriously.

The third is a simple reflection on human nature. People are naturally creative. They like to personalize things – including procedures. They love to find their own way of doing things, usually to their advantage: a quicker way, an easier way, 'my way'. This creative energy is worth using – but to improve procedures in a controlled way, not informally to modify or ignore them. Encouraging open questioning of procedures is a very healthy process. It brings out into the open

- concerns about the operability of procedures
- whether the procedure is the most effective way of controlling the process
- misunderstandings about the procedure and its application.

Whilst there are clear benefits from this approach, there are also some consequences that have to be managed – you have to operate a management system which values this approach and which will take action on those ideas which will improve current procedures. This can be difficult to achieve when the managers have written the procedures themselves, especially if procedures have been written without a clear understanding about the *practical* difficulties of implementation. Managers – and anyone else, for that matter – may well not take too kindly to questions or proposals which challenge the effectiveness of systems of work they have put in place. One way round this problem lies in the writing of the procedures in the first place, and relates to the involvement (or, rather more commonly, the exclusion) of those who are expected to follow procedures in drawing them up in the first place. Their involvement has two distinct advantages:

1. they will think through the practical implications of the procedure, helping to ensure that the procedure is not later ignored or modified because of its lack of practicability
2. there will be less resistance to challenge and change to the procedure as new best practice emerges, either from the issue of new industry or national standards, or from experience in operating the procedure 'on the ground'.

This 'ownership' of procedures is vitally important, yet it so often fails to happen. Maybe it is the way in which procedures are subconsciously regarded by us all – our feelings that they are imposed rather than the product of our own creation, that they are static rather than dynamic, that they imply a lack of trust in our individual professionalism. There are major issues here about communication, involvement, training and organizational culture – all of which demonstrate the many different ways in which procedures can fail us. Not only are there many ways in which written procedures can be deficient, but also many ways in which we can fail to translate them into practice.

Job safety analysis

A powerful but much underused technique for defining the procedural requirements of local activities is job safety analysis (JSA). This is a technique in which an operating team, with its team leader, takes time out to analyse the activity step by step. For each step of the operation the team identifies the hazards – largely drawn from the operating experience of the team members – and what options there are for minimizing the risk of injury or damage from the hazards.

The analysis may well conclude that some of the hazards cannot be overcome satisfactorily by procedures. In some cases, physical controls may be required. Solutions to other hazards may be by simply doing the job in a slightly different way. The residual hazards which can be contained by adopting a safe system of work can then be captured succinctly in a short procedure. The advantages of this technique are:

- it is done by the operating team itself, drawing on their practical experience
- it flushes out operating practices which may not otherwise come to light (custom and practice)
- it draws on the creativity of the operating team to identify solutions
- the resultant procedure has a high level of ownership with the team, and can be used as a training aid for future new starters
- the exercise doubles as a learning event, both for the team and the team leader (who often sees for the first time how things are *really* done).

Are there any disadvantages to the technique? I have never found any. It takes up relatively little time, and those taking part always seem to surprise themselves by the ingenuity of their solutions; it is a productive way of having fun!

Earlier we explored those circumstances in which procedures are needed, and where other forms of control are more appropriate. The balance is an important one. Failure to proceduralize activities where the consequences of failure are significant can lead to real problems. On the other hand, over-proceduralizing our activities runs the risk both of lack of distinction between the important and the less important, and of people simply switching off their own good judgement.

In their work, people will from time to time come across situations which are not covered by formal procedures, but where there may be some risk of injury or damage, depending on how the job is done. In these situations, what would we like people to do? Obviously, we would like them to take the 'safe' option. So how do we help to ensure that they make this preferred decision? The first is to let people know that we value their exercising their best judgement, based on their experience, and if they have any doubts about the safety of their planned action, they should consult with others who will be able to help to make an informed judgement about the safety of the approach. The second comes back to your organizational values: strong organizational values on safety make it that much easier for people to choose the safe way, even where pressure of work may drive the balance the other way. The third is education. The more you educate your people in the organizational values and practices, the more they will understand what to do to work safely, without the need to resort back to the written procedures; let us face it, how many times do people do this anyway?

Not in your back yard?

We have seen how standards and procedures are the backbone of safe performance, and we have seen, from some very significant incidents, how badly things can go wrong if they are not followed. We have also seen how easy it is to dictate against procedures working effectively. These, however, are examples from other people's organizations. Yours is, of course, quite different. Or is it? Take half an hour of your time and talk with a few people who have to operate your local procedures. Maybe you could tell them you are carrying out a straw poll on how good the procedures themselves are (focusing on the procedures rather than the conformance will remove the threat!). Ask them what they think about the procedures they are expected to operate. Ask them how many of the procedures are difficult to work with. Then, if you're sure you'll get an honest answer, ask them which procedures they do not follow, either in part or entirely. Then ask them why. If you are able to get at the truth, I can guarantee you a few surprises and, although you may not take much comfort from the outcome, at least you'll know where you stand.

Chapter 6
Training strategy

'A little learning is a dangerous thing, so it's much safer to get a lot of it'

Going back to the examples in the opening chapter, we saw how weaknesses in the arrangements for operational safety can result in major loss. In some of those incidents, there were shortcomings in the basic procedures which were written in recognition that there were operational risks which required formal controls. In most of those incidents there were clear failures in communication and training, made evident by:

- lack of clarity about the job role
- lack of awareness about job accountabilities
- lack of technical competence
- failure to learn from previous incidents.

As always, these accidents resulted from a combination of factors. However, you can see from the analysis that had adequate training been in place the chain of events would in all probability have been broken. Indeed, you can quickly come to a conclusion that failure to provide adequate training is inevitably a prime cause of incidents, for how can we expect people to work in accord with important safety standards if we don't make it clear *what* has to be done, and *why*.

There is a simple, basic cycle (Figure 6.1) that has to be working well for any chance of assurance of reasonable safety performance. Breaking this cycle disables this assurance: if there is no effective policy, the other three elements are unlikely to be strongly in place; if standards and procedures are poor, how can training be structured and auditing be of any real value; if training is not carried out effectively, people cannot deliver the standards required. . .and so on.

Where people are working with any significant hazard the penalty for failure is the increased likelihood of a major accident. Conversely, good training is an insurance against things going wrong. An alert, well trained,

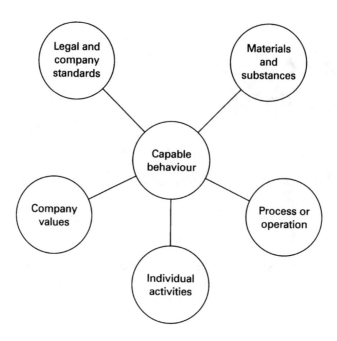

Figure 6.1

thinking individual would almost certainly have realized what was going wrong in all our headline incidents and would have intervened to prevent a catastrophe.

There are countless examples of tragedies from failure to train people in work hazards and how to avoid them. New employees, especially young people, are particularly vulnerable. Casebooks are littered with stories: a youth clearing a blockage out of a plastics' granulating machine hopper, falling into the mechanism and blocking the machine in turn with his corpse; a contractor working on a fragile roof without supporting boards and falling through the roof to his death ten metres below; operators cleaning out a vessel into which asphyxiant gas is leaking, causing their unconsciousness and eventual death. More often than not, these tragedies do not involve complex or unusual work hazards – they are commonplace and simple to guard against. But simple lack of awareness or understanding can be a killer.

We accept – even if we don't welcome – paying insurance to offset the financial effects of accidents. Yet many companies think twice about investing in a form of insurance that actually gives a return on investment – the training and development of their people. What do you gain from a financial investment? Roughly, it looks something like this:

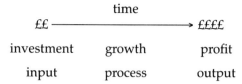

investment   growth   profit

input   process   output

Similarly, training as an investment looks something like this:

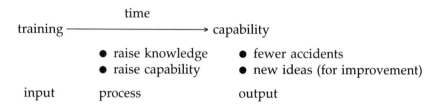

input   process   output

The better the standard of the input, the better the value of the output. Skills and knowledge are an operating asset. In any people-intensive business they are the *most* important operating asset: in tight markets, they are often our competitive edge.

Training, at its most basic, equips people with what they need to know. At its best it achieves two things. It tells people what they specifically need to know about activities relevant to them *and* it gives them an understanding of organizational values on safety that will enable them to do the right thing in situations which fall outside their scope of their experience.

It's not just in the on-going, day-to-day management of safety that training is important. Chapter 2 describes the process of generating a successful safety improvement strategy in which training is a key element. It stands in its own right as an activity specifically aimed at equipping people with what they need to know and be able to do. It is also present implicitly in the other elements: the communication of the ideas which make up policy, strategy, standards and procedures is also an activity which engages training or learning.

The development of 'organizational culture' also has strong training and learning connections. What is organizational culture if not the sum of individuals' attitudes and behaviours? We cannot modify the one without modifying the others – and vice versa.

The structure of this chapter follows the simple process illustrated in Figure 6.1, where we first decide what people need to know and be able to do, second check out what they currently know and are able to do, third provide learning activities to bridge the gap, fourth ensure that any learning is consolidated into normal work activities, and last check to see that the requirements of 'know/can do' are being expressed as normal behaviour, and give it positive reinforcement.

Achieving capable behaviour

What any manager should be seeking to achieve by putting energy and resource into training is a set of clear outcomes which describe something we might call 'capable behaviour'. Set in a safety context, these outcomes would be:

- people believe that working safely is important
- people understand what their job requires of them
- people have sufficient skill, knowledge and experience (capability) to do the job safely
- people actually do work safely
- people continue to learn from new experiences.

If people are expressing this capable behaviour, then we are home and dry! So let us look further at this concept of capable behaviour. How do we decide what someone's capable behaviour should be? One starting point might be people's job descriptions. Most people have some form of written job description which identifies their key accountabilities and responsibilities. This is often couched in terms of output requirements and not in terms of behavioural descriptions. 'Operate a transport service from warehouse to customer safely and on time', or 'Ensure the safety of employees in manufacturing facility XYZ' are not particularly helpful statements because, whilst it tells us what the ultimate target is (the output), it doesn't tell us much about what capabilities are needed to achieve that target.

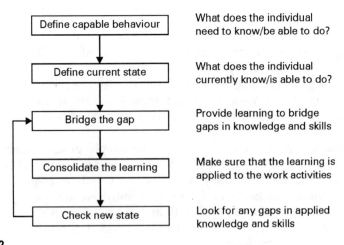

Figure 6.2

It is much more helpful to look at capable behaviour in the context of those elements which influence it. In Figure 6.2 capable behaviour is associated with five key areas of knowledge and expertise:

- mandatory requirements
- materials and substances
- processes or activities
- the individual's specific activities
- company values.

Mandatory requirements

Mandatory requirements are those which are required by law, or by company standards (expressed locally as safety procedures). Some legislation is very specific, particularly that which is directed at specific substances (asbestos, carcinogens, for example) or operations (machinery guarding, for example), and some very general – the Health and Safety at Work, etc. Act, for example. Generally, the legal requirements which are relevant to your operations will be contained within company standards, and therefore in local procedures.

Materials and substances

These are the materials and substances which are handled during operations, the use of or exposure to which could result in injury, ill health or environmental incident. This would include chemical, biological and physical agents, such as those in Table 4.1 in Chapter 4 – those which form part of any starting, intermediate and finished materials.

Process/operation

The nature of the fundamental process or operation will itself place requirements on the individual. Pressure, temperature, volume, batch or continuous operation, control systems, actions in the event of abnormality operation etc. are all possible factors for manufacturing. Factors such as travel standards, loading, fuelling, routing, etc. would be relevant to a transport operation.

Individual's specific activities

The very nature of the job role the individual is carrying out places requirements on what they need to know and be able to do. This can make a great difference to the definition of capable behaviour even within the same sphere of operation. For example, both an anaesthetist and a surgeon require to know about physiology, anatomy and surgical procedures, but the requirements on them in terms of 'can do' are very different.

Company values

Last, but not least, everybody needs to understand the values of the organization in which they are working. This gives them their 'fall back' position in situations where there is lack of clarity, compromise or conflict. These values underpin *any* situation, even where specific procedures are defined, and allow people to test for themselves how well the established ways of working align to these core values.

So, if I:

- understand what the law and company safety standards require (and why)
- know the hazards of materials and substances in my operations and how to contain them
- know the hazards specific to my own activities and how to manage them
- know what values to fall back on in new situations

I am well equipped to do my job safely.

An even more detailed understanding of capable behaviour comes from a knowledge of what *level* of proficiency is required in the *individual activity* element of the capable behaviour model (Figure 6.2). In all our jobs we need different levels of capability for different aspects of what we do. These different levels can be described as *expert* (E), *competent* (C) and *aware* (A), and should match the varying requirements of our individual roles. Let's look at a simplified, idealized example to illustrate.

A process engineer designing a chemical plant would need to be an *expert* on engineering standards and practice, would need to be *competent* at identifying what type of materials of construction would be appropriate for which chemicals, and *aware* of the practical constraints on pipefitting techniques. A materials engineer would need to be an *expert* on the effects of different chemicals on different materials of construction, would need to be *competent* in engineering standards and practice, and *aware* of pipefitting and similar engineering construction practice. A pipefitter would need to be *expert* on pipefitting and similar engineering construction practice, *competent* at identifying what materials of construction are appropriate/not appropriate for chemicals being used on his plant, and *aware* of general engineering standards.

Between the three, there are adequate overlapping competencies which help to ensure that there is no gap in competencies throughout the whole process of designing and constructing an engineered system for processing chemicals (Figure 6.3).

	Design	Moc*	Fitting
Process engineer	E	C	A
Materials engineer	C	E	A
Pipe fitter	A	C	E

*Materials of construction

Figure 6.3

Let's take another example relating to a fairly small operation which manufactures a product, one constituent of which is a flammable liquid. Everyone connected with this operation will need an awareness that the flammable substance is present, what the potential hazard is, what the risks are of ignition (and therefore what they may or may not do in terms of personal behaviour e.g. carrying smoking materials), and what to do if an escape of material or a fire actually occurs (e.g. emergency response, evacuation procedure etc.). This awareness level of capability may well extend to contractors who regularly come onto the site – they certainly need some basic awareness of what to do and not to do. And what about site visitors? They may also need a basic level of awareness to ensure they don't do anything inadvertently to put themselves, or others on the site, at risk. Awareness may go even further than this – the neighbours of the site, for example. Making sure people know about hazards, containment measures and emergency measures in the event of an incident which could affect the local community is, at the very least, good public relations, and may well be a legal responsibility depending on the nature of your operations.

The level described as 'competent' will in essence be the person displaying all the 'capable behaviours' required to operate safely in this work environment. They will certainly have all the knowledge of the 'awareness' level and in addition, for example, they will know how to operate and/or maintain the process equipment safely; they will know the control philosophy and procedure for handling the hazardous material; they will understand the flammability and explosion hazards and limits; they will be fully conversant with implementing emergency actions on the plant and they may well be able to extinguish fires and administer first aid.

The competent level implies all the knowledge and capability required to control the operation in normal circumstances and to be able to recognize when it goes out of control and be able to deal with such a situation.

The definition of the 'expert' level depends on the nature of the operation. In the example above, the expert level may well be the manager in charge of the manufacturing operation, or possibly someone responsible for overseeing the technical and/or engineering standards of the process. It is the person who understands the theoretical basis of the process, who is conversant with the design, construction, control and operating philosophy and parameters, and who can make sound judgements on the hazards and risks involved both in normal operation and where operational changes are being considered.

What we are doing by examining these aspects of capable behaviour is to define the fundamentals of what individuals need to know and be able to do to carry out their job roles safely. All jobs will have a basic content of this nature. Some will be very simple and obvious, others much more complex and will require careful analysis. However, in all cases we need to be absolutely explicit about what is required before we can move on to the next stage of defining what training needs to be delivered. If you have difficulty in starting this process, or you would like to verify your analysis, it is very instructive to look at a few 'role models'. Who are your best performers with regard to the competent level in performance? What knowledge, capability and experience does your acknowledged expert actually have?

However carefully you define the content of capable behaviour and, particularly, the expert level, it is always a good idea to check outside your organization to make sure that you really are looking at best practice. For example, who do you regard as 'best in class' in your sphere of operation? What professional qualifications and levels of operational experience do *they* require of their 'experts'? What guidance is given, for example, by Industry Associations, and Government agencies?

Identifying the 'capability gap'

Having characterized capable behaviour we can move on to identify what people *actually* know and can do. It is this gap between the requirement and the actual which specifies the training need, and this process of comparison is called *training needs analysis*.

There are many ways of carrying out this analysis, but the most simple, 'user friendly' approach is by discussion between individuals and their line managers, or by self-analysis. For new recruits, this should be done formally to ensure that key training needs are identified and important gaps bridged before the recruit is allowed to start work. Some issues are so essential to safe working that they require periodic retraining – 'refresher' training – to ensure that people's capability remains high. Examples are emergency procedures, work with unguarded parts of moving machinery, 'permit to work' issuing authorizations. These are the issues that we cannot allow people to get wrong because of the potential seriousness of the consequences.

So we identify the capabilities, the gap between required capability and actual and, having done that, we are ready to close the gap. However, before we move on to the ways in which this can be done, we will pick up a few key principles about people and learning.

Some basic principles about people and learning

So far, the process I have been describing may have seemed a bit mechanistic. That is because identifying the gap between what people need to know/be able to do and what they actually *do* know/can do is fairly straightforward. However, people are not vessels which can simply be filled with information. People are very different, with different capacities to learn, different styles of learning and different past learning experiences which make them more or less receptive to accepting new knowledge. For these reasons, there is no one single way of increasing people's capabilities through learning. Here, however, are some basic pointers or principles which should always be considered when designing learning experiences for your people.

First, people need to understand the *purpose* of the training – what the new knowledge and skills are and why they are needed. So the training has to be set in the context of the needs of the job – the skills and knowledge that are important to do the job safely and well. It is helpful for people, too, to see where they stand in relation to the required skills and knowledge, so they can assess their capability gap for themselves, so that they can personally measure progress towards the new skills and knowledge their job demands. As part of this personal journey, feedback to the individual is important to help reinforce the progress being made in a realistic and objective way, with constructive critical appraisal where this is necessary.

Second, people have to have *ownership* of their learning. Remember the points made in Chapter 2 about ownership of change – progress is enhanced when people buy into the change. Learning is no different. People are capable of learning much more when they accept and take responsibility for their own learning. At best, people not only accept their responsibility for learning the essentials, but continue to assess their capability beyond the essentials, and progressively and continuously raise their capability, often improving their operating processes as they go.

Next, people should be given the opportunity to 'personalize' their learning. At best, this means that they are able to relate the new learning to existing knowledge and past experiences, and to assimilate the new knowledge at their own pace. The way in which new knowledge is presented often has a major impact on its rate of acceptance and assimilation. If the presentation style or method fits with the learner's pattern of learning, understanding and acceptance are greatly accelerated.

The next principle is that the *climate* for learning is important. Many people have negative past experiences about learning – typically, people don't want to appear ignorant, deficient, slow (and so on) compared with others. Their new learning experiences, therefore, need to be open and unthreatening, respecting the learning styles and pace of all the learners – and in a way that generates a supportive climate both between learners in a group and between individual learners and the tutor.

Next, learning is enormously reinforced by practice. It is one thing to hear the facts or theory. It is quite another to test out the theory in practice. For this reason, any learning experience in which the facts are supplemented by discussions around application and, wherever possible, by practical application itself, will be the most successful in cementing the learning and making its application much more likely and effective.

Finally, giving people the skills to learn more effectively can be of enormous benefit in their future learning, especially self-learning. This may sound unusual, but consider – if people's capability to learn is raised (and there are techniques which can help) then all their future learning is made that much easier. This is both to their benefit and to that of the organization, since they learn quicker and are more likely to want to learn in the absence of further stimulus from others.

There are well established techniques for the principles just described, and they can be used in the design of learning events. Using these techniques helps to make the very best use of the time out for learning, and ensures a high likelihood that the learning will be effectively applied.

Types of learning

Many people think of training as an 'off-the-job' course where a trainer delivers some materials to a number of participants in a fairly formal process. This characteristic perception is often what happens in practice. However, the range of possibilities is very much wider.

First, let us substitute the term 'training' for the term 'learning activities'. We can then visualize a much more complete picture in terms of learning opportunities, e.g.:

- formal/informal
- on-the-job/off-the-job
- work-based activities/special project
- delivered by tutors in person/delivered through 'open learning packages'.

Let's take a look at these types of learning activity in some more detail.

Off-the-job training

Off-the-job training can be a very efficient learning process where there is a need to ensure a consistent handover of information and approach and where there is the opportunity to engage a good knowledgeable presenter. Moreover, where a reasonably large group of people can interact with the tutor(s) it can be very cost effective. Formal courses can also be particularly effective where concepts are being explored and evaluated and where there is an opportunity for participants to contribute, and to learn from each other's individual experiences.

Where the outcomes are less prescribed and debate is being sought, this type of training becomes more of a 'workshop'. An example of this would be the generation of safety 'vision' by a management group, or the generation of an auditing plan by representative members of an operating unit.

On-the-job training

Learning 'on-the-job' can be a very effective process. The learning can be made to be very timely, can be easily reinforced by example and trial by the learner under supervised conditions, and is cost effective – no time away from the job.

Unfortunately, however, this too often is the default mode when insufficient thought has been given to training and people learn by 'sitting with Nellie' – an approach which guarantees that all of the misconceptions, bad habits and unofficial work practices are efficiently transferred to the novice. It is much more effective to set out a progressive programme based on the concepts of 'capable behaviour' and the real needs of the individual concerned. In this way specific activities can be planned into the work schedule, the individual can have procedures demonstrated to him or her, notes can be taken and then he or she can perform the work activity in closely supervised conditions with coaching and feedback built into the activity. This is particularly important when considering the timing of training in relation to the need for it. Training is most effective when the individual has an immediate opportunity to apply and reinforce the learning in a real situation.

Where a person's knowledge or capability is being deliberately extended, perhaps in anticipation of a job change or for personal development reasons, specifically designed work-based projects may be set up where the individual spends some allocated time investigating aspects of the current work activities or of activities in a related job role. He or she may spend time with colleagues or supervisors in other areas or perhaps carry out a study on a particular aspect of the current work operations.

The whole idea of learning on-the-job and from the job is very well practised. The very notions of apprenticeships for tradesmen and medical students spending time as housemen in hospitals before entering general practice are examples of this approach. The critical issue here is that it is a planned experience with a purpose and an agenda of issues that will be covered. Another key factor is the support and guidance the novice receives from the experienced tutor. We will explore these tutoring roles further later in this chapter.

Open learning

Open learning (sometimes known as 'distance learning') is a relatively new concept, although some elements of it (e.g. texts from which one can learn) are as old as the hills. Open learning basically puts the learner in control of the learning process. The relevant information can be made available in a variety of media and the individual chooses when, how, how fast and sometimes even where he or she wants to absorb it.

There is a great deal of excellent material available in open learning format. Additions are being made all the time to public, commercially available, open learning packages, and it is often cost effective to assemble organization-specific or job-specific packages. The great advantage for the training provider is that once the initial cost of the materials has been made it can be used again and again at little additional cost. Open learning is most appropriate where you want solely to transfer knowledge and facts. There are some interactive packages where concepts and behaviours can be explored but this often requires further tutorial support in order for individuals to test out and apply the learning in practice. It is best suited to individual learning where people work essentially alone, interfacing with the material at their own pace and at a time (and possibly location) of their own choosing.

Here is a little additional information about open learning materials.

Texts
While books could be considered as open learning, the essential interactive process is not normally built into the typical textbook. Carefully designed authoring is needed to provide an interactive workbook where the recipient's grasp of knowledge and its application can be tested out during the learning process. It is particularly important that the really critical pieces of information are focused and implanted during the learning process – ordinary text books do not usually bring these factors sufficiently far forward into the learner's consciousness. Advantages of open learning work books are that they are cheap, portable, easy to use and relatively easy to construct.

Audio tapes
Simple audio cassettes can be used to support formal courses and texts by providing summaries and reminders of key information or approaches. Probably less effective as a primary learning tool, audio tapes are none the

less cheap, very portable (e.g. can be listened to in the car, at home etc.) and reusable. They can be used to considerable effect for retraining, where the required knowledge and skills have already been given, and the need for interaction is low, other than in evaluation of the required capabilities. Take care when selecting audio tapes for training; some are very good, some are pretty awful.

Video tapes

Video tapes are excellent where situational factors need to be demonstrated and role play is an important consideration.They are often accompanied by some form of workbook which enables the student to replay parts of the tape in order to check out key issues and deepen his or her understanding. Drawbacks are high initial cost (especially if preparing a locally specific video) and a tendency to get out of date. Advantages are almost infinite replayability, and strong appeal to the visual sense which is an important learning aid for most people.

Computer-based training (CBT)

CBT is software-based training in the form of packages which usually run on personal computers. These can be simple (mimicking open learning texts) or more sophisticated – perhaps building in quite complex databases and options to test and evaluate the learner's progress. In the extreme one could consider using 'expert systems' where the software package has taken the place of the expert in terms of the range and depth of knowledge on a topic and the normal operating algorithms required to use that knowledge satisfactorily.

CBT's great advantage is its flexibility. Its biggest drawback is its cost, and the learner needs to have access to a suitable personal computer. However, very little computer skills are normally needed, since CBT packages are normally constructed to be very user-friendly.

Interactive video

Interactive videos have the video sequences encoded onto a laser disk. This means that choices of sequence can be made by the student enabling 'right' and 'wrong' selection and their outcomes to be displayed. This is a very powerful learning tool, and can be set up to evaluate the learner's selection, and therefore to track his or her learning progress. The great disadvantage is the very high cost, both for the disks and for the laser readers.

Multimedia

The increasing sophistication of personal computers is now enabling a multimedia approach to open learning to be developed. This is where an interactive computer program can access text, audio (including speech and music) and any form of visual data (photographs, video sequences, graphics). The potential of these packages is enormous, and the overall editing software is becoming more and more easy to use. Again, the major drawback of this tool is its cost, although this is reducing in real terms as the programs get smarter and more software is available on the market.

Simulation

In extreme cases, where the hazards are very large or the risks cannot be completely eliminated, the work environment (or some aspects of it) can be simulated so that the learner can learn experientially without any detrimental consequences. The classic example of this approach is in airline pilot training where the trainee can experience all conceivable situations of normal and abnormal operation without leaving the ground. Only when 'capable behaviour' has been demonstrated satisfactorily in the simulator is the pilot allowed to fly the real thing.

Clearly, the expense of building such a complex simulator is extremely high. However, even in quite simple operations the use of models, both physical and in computer program form can be a helpful approach in getting the trainee to think through different operational situations.

How we learn

Earlier, we looked at some basic principles about people and learning. One of the points made was that people learn in different ways, and that the particular approach to learning which an individual prefers has a strong influence on the successful take up of learning from any learning opportunity. For this reason, care must be taken by those preparing learning activities not to assume that all learners will find a particular approach equally useful, nor that the author's own preferred learning style should be the only approach adopted.

Honey and Mumford categorized four distinct learning styles, which form a helpful framework for training providers to consider when assembling their material and designing the learning activity. The four styles are:

● activist
● reflector
● theorist
● pragmatist.

Let's look briefly at each one in turn.

People with a preference for the *activist* style learn well from new experiences. They like to get 'stuck in' and have a go at challenging tasks with few or no constraints in terms of structure or boundaries. They like to be in the centre of things and to be engaging with others. They find difficulties with situations where they have to be passive (e.g. listening to lectures, reading or even watching others do things), do not enjoy following procedures or instructions which they feel inhibit them, and will avoid solitary activities.

Those who prefer a *reflector* style learn well from situations where they can plan an activity, observe what happens and then review and analyse the outcomes. They prefer structured learning experiences which are 'safe' in terms of pressure on timescales and achievement needs. They will not enjoy being put in situations where they have to 'role play', where no planning time is allowed before action, or where they feel there is not enough information available.

Learners with a *theorist* style like to have concepts and models to which to relate their learning. They choose to explore the relationships between new information and basic theories and are able to analyse complex data and situations using elegant logic. Situations where ideas are presented only superficially, or there appear to be contradictions in approach will not be helpful to theorists who like an intellectually soundly-based methodology with good supporting data and statistics.

Individuals with a *pragmatist* style are always seeking real outcomes from what they learn, so anything which can be used immediately to help them to achieve their work goals better will be seized upon. They respect practical experts and respond well to coaching. Examples and demonstration of how the learning has been or could be used are always welcome. This essentially practical style will find difficulty with learning activities structured in ways which don't make the benefits obvious, where the ideas or presenters are highly theoretical or where guidelines in how to go about doing things are not explicit.

The learning style preference of each individual will vary and the degree of exclusive preference to one style is also a factor. These profiles can be measured using Honey and Mumford's 'learning styles' questionnaires, but it is unusual (and probably unrealistic) to expect to tailor learning activities specifically to a single individual's learning profile.

More important is for the learning activity designer to check whether all four of the learning styles have been considered in the design, and whether there is sufficient appeal within the offering to capture the interest of each one of the categories who may well be represented in the audience.

Design of learning materials

It is not only the design of the learning material, however, which has to be considered. The very nature of the learning environment can be important too. Learners may have very varied and different requirements with regard to the structure of the learning environment, for example, whether the learning is clearly linked to an explicit outcome or validation, or whether there is there an opportunity to put the learning into practice.

Other learners may require learning which is structured in outline, but less detailed in its texture. An example is a regular timetabled 'tutorial' briefing/report with tasks and reading/study assignments in between.

Yet others may prefer a truly flexible approach to structuring their learning, simply taking opportunities to pick up information and skills as they present themselves naturally during work (or even outside work). This type of learner may be continuously upgrading his/her knowledge and capability by examining similarities and differences between all their experiences.

There are social considerations in setting up the learning environment, too. Some learners prefer to work alone, others require a close one to one support from a 'tutor', and others are much happier learning in a group situation, exploring and testing ideas with colleagues and peers.

Sensory preferences can also influence learning capability

Which of the senses should the learning material appeal to? Most individuals respond well to material which impacts them visually ('a picture paints a thousand words'). So diagrams, photographs, cartoons, graphs, charts and the use of colour all help to get the key messages across. Other people are more aurally than visually orientated, when tone, use of specific words and other sounds (music, laughter etc.) can have a strong influence on the learning. There are other learners who respond strongly kinaesthetically (i.e. by sense of smell, touch and feel). For these people the physical engagement with the equipment or components, how fitting them together actually feels, or the very atmosphere generated in the learning area can be important factors in their take up of learning.

Who should do the training?

Involvement in training is very much a 'horses for courses' affair. In open learning, the learner primarily takes the initiative, although it is always helpful to have an experienced mentor available in support of the learning – for example, where the learner has questions which arise from the open learning material. In other cases a straight 'teaching' style using an experienced trainer may be appropriate. In other cases, where a straight teaching style is not in itself sufficient in terms of understanding the material fully, the learning may better be catalysed by a tutor who can expand and challenge the learner's insight.

If it is an activity where practice and a high level of skill are necessary the provision of a coach may be desirable. Here the role of the coach is to demonstrate the activity, provide a 'safe' environment for the learner to practise and then to give positive feedback on the performance of the learner to encourage success while correcting errors and deficiencies. An ideal situation is where the individual's manager fulfils this role. The involvement of the manager is particularly useful in ensuring that the manager understands the issue – and often learns a good deal in the process. In addition, the identification of a role model is a most valuable resource for a learner to have – someone whose 'normal' performance encompasses and exhibits all that the learner is striving to master can be an inspiring example.

Other considerations

All the factors we have explored so far can help the learning event designer to consider the most appropriate form of learning material for individual learners. The critical point is that the learning event designer must appeal to as many of these requirements as possible, with the aim of minimizing any blockages which may prevent people learning – in whatever form it happens to be. Nevertheless, there are still some potentially fatal obstacles which must be overcome to ensure success for our learners. While you may not be directly or personally accountable, the motivation of the learners is critical. If the learner has been 'instructed' to attend, or does not see the value in learning about a particular topic then he or she is not likely to approach the learning activity receptively.

Similarly, time must be made available. This may imply arranging cover to release the person or reallocating or rescheduling work. If the training is important then it must be perceived as such. Cramming it in over a lunch break or forcing people to pile work up while they attend a course is likely to be counter-productive.

Consolidation of learning into practice

So far we have not considered the actual content of safety learning activities, other than in a generic sense. Without being complete or prescriptive, a typical list of available training would include topics such as:

- induction, including safety values and culture
- company standards and local procedures
- hazard assessment techniques
- designing safe systems of work
- emergency procedures
- incident investigation
- safety auditing.

Whatever the topic and content of the learning event it is vital that the learner has an immediate opportunity to try it out on the job and in the workplace. Here the role of the supervisor or manager as a coach is paramount. The individual needs to be able to discuss the learning and consolidate it into the normal work processes. The coach must allow this to happen for the learning to be fully internalized and become an instinctive part of the person's behaviour.

In some instances this may just not be possible. Some safety training is aimed at circumstances which do not normally occur – for example fire fighting or emergency evacuation procedures. In these cases it is essential that regular refresher training or practice sessions are set up so that the learners have opportunities to engage with the event with sufficient frequency to establish proficiency.

Evaluating the learning

Now we will turn to evaluation of the learning – in other words, how do we know the effort we have put in has been rewarded by a change in our learner's capability? Let us look at the options. Perhaps the most obvious, and certainly one we are all used to, is to test the learner by asking questions – a written or oral examination. This is a good process if we are checking to see whether our learner has acquired knowledge.

It is much less valid if we are checking on skill or 'can do' capability. For example, just because someone knows the contents of the highway code, it does not mean that they are a proficient driver. (Not only that, but they may have forgotten these facts by next week or next month.) For checking skill and 'can-do' capability it is much more useful to assess the learner in an on-the-job situation. Are they performing the task accurately and exhibiting the specific skills and behaviours with regard to safety that have been taught or demonstrated? This assessment role is often performed by the individual's line manager, who should have a well thought through set of criteria and performance standards by which to make an objective measurement.

In some cases (e.g. where public safety interests are involved) this measurement may be carried out by an external assessor. As described earlier in this chapter, in extreme cases (e.g. airline pilot training) this assessment is made on a simulator before the learner is allowed to carry out the procedures in the normal operating environment. For non-critical situations, self-assessment may be possible, where the learner can use a simple questionnaire to check out his or her capability. This approach is clearly not appropriate where there are any aspects of operation which may present a significant hazard – either to the learner or anyone else.

Let us assume that, whatever method we have chosen to evaluate it, our learner has successfully internalized and consolidated the new capability and has demonstrated it satisfactorily in the workplace. What should we do next? Well, at the very least he or she should be given some positive feedback on the achievement, which should always be recorded in a personal development record.

We have now completed the five stages of the model in Figure 6.1. We have:

- identified the training needs
- defined the current state of knowledge and skills
- identified the gap in skills and knowledge and provided learning activities to bridge the gap
- consolidated the learning into practice, and
- evaluated the learning.

Finished? Well, not quite. Like any learning process, these stages really form an overall cycle of:

plan ... do ... review ... improve

We are not just looking for compliance. We want to see people develop their capability to contribute proactively towards making their workplace safer and more productive. The end goal of training is ultimately the commitment, competence and delivery of continuous improvement.

Chapter 7
Auditing

'Auditing is a flexible tool: it can be used by oneself as a means of self improvement and by others as a means of corporate vengeance'

The purpose of auditing

Auditing is usually associated with financial auditing, and that is actually a good starting point to understand why we need to do safety auditing. So why do we do financial auditing? We do it essentially to check whether the financial arrangements and transactions are properly accounted for, and as a 'financial health check' of the business. Business financial operations are complex, and a business needs to keep closely in touch with how it is managing its finances. Problems with cash flow or income and expenditure can result in business collapse.

Just the same principles apply to the safety management arrangements of a business. The safety systems by which a business operates can be equally complex as the financial arrangements, and failures can be equally damaging. So checks on the operation of safety management systems are important.

Auditing is generally regarded as a means of checking that procedural arrangements are being followed. This is an important aspect of auditing; procedures are the way in which we describe the best current practice for the activity and the more consistently those procedures are followed, the more confident we can be that unwanted incidents will not occur. There is another important aspect of auditing, relating to the development of best current practice. If we identify the best current practice *today*, using today's knowledge, can we assume that it will be best current practice in a year's time? Of course not. For this reason, auditing of safety management systems should include an element of systems improvement. It is this aspect of auditing which ensures that changes – in legislation, in industry or commerce standards, in internal standards or any other developments in good practice – can be applied in updating and improving safety systems.

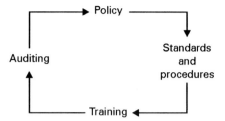

Figure 7.1

Auditing is an essential part of the means of assuring safe operation through the cycle described in Figure 7.1, through which policy, standards and procedures can be continuously monitored and improved to reflect best current practice. Without auditing, the chain is broken and there can be no way of knowing whether best practice standards and procedures are being applied, and no means of improving policy, standards and procedures in the light of new best practice.

Structured auditing

Comprehensive safety auditing arrangements are those which are capable of answering the following questions:

- how well do people understand local procedures?
- how well are we conforming with our local procedures?
- how relevant are our local procedures to local circumstances?
- how closely do local procedures comply with company standards?
- how comprehensive are our safety management arrangements, and how well are they working?

This spread of questions is very broad, and is targeted at different levels of operation. At the delivery level, *operational* auditing (see Figure 7.2) addresses the first three questions:

- how well do people understand local procedures?
- how well are we conforming with our local procedures?
- how relevant are our local procedures to local circumstances?

The third question needs to be asked, especially where there are differences between the requirements of the procedure and the way in which the work is *actually* done. In these circumstances it is not enough just to note non-conformance – we should ask whether the procedure is suitable for the local circumstances. For example, a procedure for work in a service cavity which requires a working platform to be constructed was not followed because it was not possible in most cases to construct such a platform inside the cavity

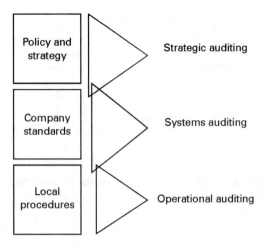

Figure 7.2

because of the configuration of service pipework. The person who had written the procedure simply hadn't taken the practical problems into account, so the requirement was ignored.

At the next level of auditing – *systems* auditing – the following questions are addressed:

- how closely do local procedures comply with company standards?
- are there any other relevant standards (e.g. national or international standards or legislation) which need to be considered?
- how well is operational auditing carried out?

This series of questions helps to identify any improvements that are needed in local instructions, in company standards (where, for example, national or international standards have changed) or in the way in which operational auditing is carried out.

At the third level, *strategic* auditing, the way in which company values and policies are made to work is examined. This looks closely at the attitudes of the senior management to their management of safety, the plans that are in place for managing safety – the arrangements for local instructions based on company standards, training, auditing, etc. – the way in which those plans are being delivered and their progress monitored, and the overall performance of the unit being audited.

These three layers of auditing represent a comprehensive check of the effectiveness of the safety management arrangements in an organization.

How is this done in practice? How manageable is such a comprehensive system? What are the key factors in making it work well without it taking up unreasonable amounts of resource? These are very practical questions, which we will examine for each of the three levels of auditing.

Operational auditing

Arguably, this is the most important layer of auditing, because it identifies any fundamental problems in the translation of company safety standards into on-the-ground operation. It is also the level which usually takes up the greatest amount of time and effort.

There are some guiding features of good operational auditing. They are:

- get everyone involved and train them to audit effectively
- audit all the relevant safety procedures, but on a prioritized basis
- adopt a consistent approach to managing the audit process
- have a robust action and review mechanism for dealing with problems.

Getting everyone involved

Getting everyone involved in auditing sounds like a tall order. So, what are the advantages? The first is that it spreads out the auditing workload, so that operational auditing does not become an unreasonable burden for a small number of people. The second is that as people learn the skills of auditing they become more aware and critical of the safety of their own activities. The third is that people at the operating level know what goes on and, providing you have an open culture, they will identify very easily any differences between 'theory' and 'practice' – a vital part of what operational auditing is about. The fourth is the involvement itself: this is an excellent and productive way of engaging everyone in the safety improvement effort.

There are some skills which need to be learned to audit effectively – not just learning the procedure that you adopt to carry out and report audits, but the ways in which to approach people and ask questions. Asking 'do you follow the procedure?' is likely to be met by a brief 'yes'! Asking 'tell me/show me how you do this' will tell you a lot more about how closely the procedure is actually followed. So, simple training in auditing skills is important.

Audit programme

Like all other systems, auditing has to be managed. An obvious way to do this is by devising an audit programme which sets out which procedures will be audited, when they will be audited, and by whom. The programme must be seen to have management support and involvement, and the programme itself must have an owner – someone responsible for monitoring its implementation, for ensuring that actions from audits are carried out, and for sharing the learning from audits within the operating unit – an audit programme manager.

A good audit programme will include all the procedures relevant to the operating unit, setting a frequency of auditing which relates both to the relative importance of each procedure and any known levels of compliance. So, for example, a safety-critical procedure which is known not to be working well should be audited – initially at least – fairly frequently, whilst a non-critical procedure which is known to be working well would receive minimum attention.

Just to get a feel for the amount of effort needed in your audit programme, if you have 100 procedures to audit, and your audit programme runs over a two-year period and, on average, each procedure is audited twice in that period, that gives you 100 audits a year. If you have 200 people in the operating unit, and they audit in pairs, then each person will carry out one audit a year – about two hours a person a year – hardly a major imposition! Even if you wished to start your auditing programme more rigorously – say, at twice this frequency – that would only commit each person to four hours of auditing a year.

Audit programmes need to be flexible so that they may be adjusted to reflect the outcome of audits. So, for example, if a frequently audited procedure is showing high levels of conformance, there is a good case for reducing the frequency of its auditing. Conversely, if an audit demonstrates significant non-conformance to a procedure, a follow-up audit may be necessary soon after corrective action has been taken to check on improvement.

Adopt a consistent approach to managing the audit process

One significant advantage of auditing is that it can give you a real indication of your *intrinsic* safety performance – the fewer non-conformances identified during auditing, the less likelihood of incidents. So if you adopt some simple ranking of compliance, you will be able to see where your strengths and weaknesse lie, and you will be able to monitor improvement.

To take advantage of this opportunity of a measurement of performance and progress, you need an auditing methodology which gives you reasonable consistency of approach. This can simply be achieved by having a guide or procedure which describes the key features for consistent and effective auditing, which can then be used in training auditors.

Here are some pointers in terms of style and approach:

● explain the purpose of auditing – to help identify improvement in operating systems, not to catch people out

- use simple language
- ask direct, open questions
- where necessary, clarify using additional questions
- ask to explain/show
- listen!
- be objective
- clarify any inconsistencies
- be patient but resolute
- verify subjective evidence
- identify and discuss non-compliance openly, in a non-threatening way
- summarize at the end, picking out strengths first
- say thanks.

A check list of questions to ask, and things to observe, is invaluable for effective and consistent auditing. If safety procedures are written well (see Chapter 5) it is easy to identify those things to examine to identify compliance. It is even more helpful if an audit check list is included with each procedure, although it should be made clear that such a check list is a basic 'starter for 10', rather than being comprehensive. It's important that auditors are giving some creative thought to their auditing. A standard audit check list for each procedure acts at least as a minimum, offering some basic questions that should not be overlooked.

Remember the essential objectives of operational audits: they are used to answer the questions:

- how well do people understand local procedures?
- how well are we conforming to our local procedures?
- how relevant are our local procedures to local circumstances?

The first of these means that auditors should ask open questions which test understanding. 'Tell me what you do' and 'Tell me how this procedure works' are two simple but effective questions which start to give you some insight into what the auditee *actually* does, what he/she understands of the procedure, and whether there are any significant differences between the requirements of the procedure and what is actually done.

It is always worth asking the question 'Which parts of the procedure are difficult, or a nuisance to follow?' or 'Which parts of the procedure would you change, and why?' These questions will often identify those parts of the procedure which are most likely not to be followed, even though an auditee might not admit to it openly. It allows you, as a follow-up to the audit, to target the problem – either by altering the procedure to make it more operable or, if the procedure really *shouldn't* be changed, to ensure that the operators understand why it *has* to be done that way. Whatever the outcome, this is a process which is most effectively done by discussion *with* the operators.

In addition to questioning operators, auditors should observe what is actually going on, to identify how well procedures are being followed. It is sometimes difficult to achieve this, especially if the activity is not carried out continuously or frequently, but it is often possible either to go to a location where the activity is most likely to be taking place, or to look for objective evidence of the procedure having been followed. Examples of this are looking at paperwork associated with the planning, carrying out and completion of a job (for example, a job controlled by a 'permit to work') and by looking at the physical state of those things which are involved in the activity – the finished job, and any personal protective equipment, and so on. Ask yourself how the work *should* have been carried out and what evidence you would expect to see of this.

As a final question to operators, it is always worth asking whether they think the procedure could be improved – either to make it easier (we picked this up earlier) or to make it safer. Remember that those who operate procedures are in the best position to judge them – they are the experts on application. It would be a waste not to use their expertise to improve the procedure, whenever possible!

Sometimes auditing raises issues which call into question the practicality of the standard upon which the local procedure is based. When this arises, the standard should be challenged – it is a way of helping to make sure the standard is progressively improved.

Recording information from the audit is important, but can be a difficult and sometimes sensitive issue. Imagine you are an operator and you are being audited by someone asking questions and writing on a clipboard as you reply. You may feel a bit threatened by this approach: 'will my answers be used in evidence against me?' Obviously, a lot depends on your company culture, and on the style of the auditor, but there are some tips worth considering. The first is to consider how many auditors you use for each operational audit. This is usually one or two people – any more than two becomes a bit overwhelming for the auditee, and unnecessarily demanding on people's time. Using two auditors, there is the advantage that one may concentrate on asking questions and talking with the auditees without the distraction of writing, and the other may stand back and make notes. Using one auditor, an effective approach is to have a good check list, minimize note taking during questioning, and make summary notes of the key points after the discussion, before moving on to the next auditee.

Measurement of compliance can be very simple, and is a very effective means of identifying areas of strengths and weakness, and of tracking progress. A common approach is a ranking of 0–5, or 0–10; for example, a 0–5 ranking may look something like this:

Ranking	Characteristic	% compliance
0	virtually zero compliance – immediate action needed	0–10
1	extensive non-compliance – priority for remedial action	10–25
2	significant non-compliance – schedule for action and re-auditing soon	25–50
3	reasonable compliance – focus improvement on the most significant issues	50–75
4	major compliance – relatively minor corrections needed	75–90
5	complete or near complete compliance – very minor improvements only	90–100

Action and review

Having a relatively simple scoring system like this helps you to decide priorities for action and, for very low scoring audits, where follow-up auditing needs to be brought forward. Following up the actions from audits, especially where the non-compliance is significant, is essential.

Failure to act quickly signals a lack of commitment to the audit process, and to the management of safety as a whole. To help ensure satisfactory progress on actions from audits within each operating unit, the audit programme manager should review progress periodically and report on progress to the operating unit manager, who is responsible for reviews with the management team. Reviews should cover:

● progress of the audit programme to schedule
● the outputs of audits – areas of strength and weakness
● progress on actions from audits, especially where audit scores are low
● trends in the output of audits
● any rescheduling of the audit programme taking into account prioritiies for follow-up auditing where audit output is very poor.

The management team must own any significant issues coming out of analysis of the audit results.

Analysis of the output data becomes increasingly difficult as the number of audit procedures and the number of audits carried out increases. The picture is further complicated if output from different operating units needs to be aggregated to give a picture across the company. In these circumstances it is worth making a small investment of a simple computer program to collect and aggregate data, and to be able to produce simple progress charts from the data. Relevant data for an operating unit would be:

- percentage of audits carried out to programme
- percentage of audit output actions carried out by agreed date
- percentage of audits in each of the six rankings 0–5
- summary of procedures gaining less than a specified audit ranking.

This data displayed, for example on a quarterly basis, would give a good indication of how well the audit programme is being implemented, and where the main weaknesses lie. Be careful about assuming trends, however, differences in overall audit output conformances from one quarter to another will not necessarily indicate shifts in overall operational compliance, since different procedures will be audited quarter by quarter.

Systems auditing

The second level of auditing is systems auditing, designed to check that local procedures are consistent with best current standards, and to review the quality of operational auditing. These are important issues, and are complementary to the objectives of operational auditing. Aside from the difference in objectives between these two types of audit, there are important *practical* differences (see Table 7.1)

Table 7.1

	Operational auditing	Systems auditing
Frequency	1 month –2 years	2–5 years
Time taken (approx)	1–2 hours	4–8 hours
No of auditors	1 or 2	2–4
Level of specialized knowledge needed	Low	High
Percentage of workforce involved	Up to 100 percent	Typically no more than 20 percent
Scope	Usually within operating units	Within or across operating units

As with operational auditing, systems auditing should be carried out to a forward schedule which reflects your current view of priority needs for auditing, and should have an owner – the systems auditing programme manager. Procedures which are particularly crucial to safe operation, and which have not been reviewed and updated for some time should be placed early in the schedule. The frequency of systems auditing of procedures may vary, depending on the number of procedures and the rate of change of internal and external standards, but a typical systems auditing programme would aim to review each procedure at least every five years. More frequent review of individual procedures would be prompted, for example, by changes in internal/external standards or legislation, or by incidents which demonstrate a shortcoming in the standard on which the procedure is based.

A key feature of operational auditing is wide involvement across the workforce, where the essential skills needed are the techniqes of auditing. Systems auditing is normally carried out by teams of managers and specialists, who have a combination of auditing skills and specialist skills in the subject to which the procedure relates. Another feature of specialist auditing is an *independent* view, with at least one of the auditors coming from outside the operating unit. So the ingredients of systems auditing are:

- unit management representation
- an independent auditor
- a specialist in the subject
- experience of systems auditing.

The most economic combination is two auditors: a unit management representative with auditing training and experience, and an independent specialist, preferably with systems auditing experience. This combination of skills may not be readily avilable at the launch of your systems auditing programme, so you may have to start with slightly larger audit teams, and work towards the more economic approach as management and specialist auditors gain experience.

Systems auditing tends to take longer than operational auditing. For the latter, one to two hours is usually enough to get a good feel for compliance levels from a reasonable sample of auditees working from a simple, pre-prepared checklist. Systems auditing requires more preparation, takes longer to carry out, and normally takes longer to prepare the audit report. Preparation for a systems audit requires the audit team to determine the current standards which apply to the procedure. This is where the specialist's knowledge comes in – he or she should be aware of both internal and external standards which apply, of any impending changes in standards, and of any relevant recent incidents which may have a bearing on the adequacy of current standards.

With all this information, the procedure is checked against the most current applicable standards, and any changes needed to the procedure are noted. Occasionally, this process identifies shortcomings in the parent standard, which should be passed on to to the person responsible for updating the standard. Where standards require particular local expertise, auditors should check to see that they are adequately qualified and trained.

The second part of the audit is a review of a sample of the reports of operational audits of the procedure, both to assess the quality of operational audits (how searching and comprehensive have the audits been, and how well is the evidence of compliance or non-compliance identified) and whether there are any trends of non-compliance which give any cause for concern.

The audit team should agree actions and dates for completion with those who will be responsible for taking action and write up the audit report to include:

- the title, scope and date of the audit
- the auditors
- consistency of the local procedure with the parent standard, and any actions
- effectiveness of operational auditing and completion of audit actions
- any shortcomings in the parent standard
- a summary of agreed actions and dates for completion.

The report should be received by those responsible for actions, and by the systems auditing programme manager, who will keep a check on the progress of the actions to completion. Just as with operational auditing, the output and progress of the systems auditing programme should be reviewed systematically by the senior manager responsible for the area covered by the scope of the systems auditing programme. The scope will depend on the range of application of the procedure, and possibly the geography of the operating units. For example, if the procedures apply across a site accommodating a number of operating units, it is economical and practical to carry out systems audits *across* operating units, with the associated advantage of sharing resources.

Strategic auditing

Strategic auditing is the third level of auditing, examining the way in which company values and policies are made to work. A strategic audit is an independent assessment which may be carried out by internal company auditing specialists, or by experienced external auditors, and is typically carried out every 2–3 years. It is essentially a 'health check' of safety management arrangements, identifying strengths and weaknesses, and giving helpful recommendations for ways of improving management of safety.

A typical strategic audit is carried out by one or more auditors over a period of one to five days, depending on the size and complexity of the site or business or function being audited. In advance of the audit, the auditors will look at the organization chart, the safety policy, the safety improvement plan, the accident record over the past few years and any other measures of performance and improvement and, if there has been a previous strategic audit, a record of progress on actions from its recommendations.

The audit would typically start with an interview with the most senior manager to discuss aims and values and overall philosophy on the management of safety, how they are translated into policy, and how that policy defines the practical arrangements for the management of safety, and the responsibilities for making the arrangements effective. The way in which these arrangements are structured into an improvement plan is examined, both against the background of the incident record over the past few years and the main hazards associated with the operations. General arrangements for the production of procedures from company safety standards, for training, for auditing and for review of progress to plans would normally also be discussed.

Interviews with members of the most senior management team would follow, again exploring aims and values and the way in which they are translated into local practice and improvement plans.

As part of the audit, the key features of safety management policy and arrangements for making it work are explored. These would include the state of progress in the translation of company safety standards into local procedures; the arrangements for training people in those procedures which are relevant to them; the arrangements for (and output from) safety auditing; analysis of the causes of incidents and any conclusions and actions resulting; and any initiatives for sustaining and improving organizational safety culture.

The aim of the auditors is to test the comprehensiveness and commitment to the arrangements for improving the management of safety, sampling for objective evidence of follow-through action on the ground without going into too much fine detail.

The outcome of a strategic audit is a factual account of the auditors' findings, drawing out the strengths in the arrangements for managing safety and identifying areas where greater emphasis is needed in support of the improvement effort. The auditors should discuss their conclusions with the most senior manager of the unit audited to clear up any questions or differences in perception before the audit report is issued.

A well run strategic audit will always be of benefit to the unit being audited. It is an opportunity to have the current position assessed independently and professionally by auditors who have a lot of experience of good practice elsewhere. The best auditors share this experience to spread good practice and to raise expectations.

No auditing, no information

Auditing is an essential if you want to keep your finger on the pulse of your safety management arrangements. We can all have our 'feel' for how things are going, but we know how misleading those 'feelings' can be without hard evidence. Being told things are working is *never* enough – nobody likes to give bad news, and there is *always* some bad news. Avoiding the bad news is the road to disaster. No-one can manage without hard facts. Good auditing gives you just that – and with the engagement and education of everyone – a powerful combination!

If we look back at the major disasters in Chapter 1, it's easy to see that in most cases no structured or serious auditing was taking place, and that the root causes of the incidents would have readily been exposed by auditing. The Piper Alpha incident, where safety auditing was carried out, demonstrates how auditing which is ill-timed or which is poorly structured, or which is carried out by poor auditors, or where shortcomings identified by audits are not taken seriously, is inevitably ineffective.

Problems are rarely difficult to find through good auditing – they almost jump out at you. Some failures are so obvious that auditing isn't necessary to identify them – the failures that led to the *Herald of Free Enterprise* and the Space Shuttle *Challenger* disaster, for example. These cases again bring home the absolutely essential point that if problems which are identified – whether through formal auditing or from simple observation – are not taken seriously and action taken, there is little point in carrying out auditing in the first place. And full steam ahead for the rocks!

Chapter 8
Measuring progress and performance

'Lies, damn lies and statistics'

Measurement is the only objective test of performance and progress. Yet how well do we use measurement to gauge them? Let's test.

List your current measures below (Figure 8.1). Then, in the middle column, describe what these measures tell you about how to improve your performance. Then, in the right hand column, take a view about how effective these measures are in determining where to focus your improvement effort: VE for very effective, E for effective, NVE for not very effective and NE for not effective at all.

Measure	What the measure tells you	Effectiveness
•		
•		
•		
•		
•		

Figure 8.1

How did you do? Well, full marks if you scored VE or E in the right hand column. If you scored NVEs or NEs, don't despair, you are not alone. Why is it that most of the frequently used measures of performance give us so little help in understanding our performance and where our strengths and weaknesses lie?

The fact is that almost all organizations measure *output* performance, usually in terms of reportable accidents. How does this help? It tells us what levels of incidents we had last week/month/year, and it adds to the pattern of historical data, and it helps us to compare ourselves with other organizations which make similar measures. However, it will not give you any indication of what your performance will be next week/month/year. Nor will it give you any indication of how close you are to a major incident because, as we saw in Chapter 1, the underlying causes of major incidents – poor policy, deficient standards and procedures, poor leadership, inadequate training, no effective auditing, and so on – do not necessarily relate well to conventional output performance measures.

It is quite all right to measure outputs such as injury rates or conformance to environmental consent levels, since they enable us to gauge our peformance against other organizations – but using them as exclusive measures of performance is inadequate. Why? For three main reasons.

● An output (reportable injury rate, for example) is the result of a combination of *many* inputs and processes, the strengths and weaknesses of which are not indicated by the output measure.
● Output is subject to natural variation: if you did *nothing* different the output would *still* vary through time.
● Exclusive focus on output measures encourages a mixture of complacency and knee-jerk reaction.

Let's look at each of these three more closely.

Outputs are the result of *many* inputs and processes

If we look again at the process model (Figure 8.2), and the key SHE factors associated with inputs, processes and outputs, we see that the outputs are dependent on those inputs and processes.

So, if we are to understand what the outputs are telling us about our performance, we have to understand better the state of those inputs and processes – how well, or otherwise, are they working, and how do they need to improve? We can only improve our output performance by improving on those inputs and processes. Once we understand this, we can set out to make improvements which we can *guarantee* will have an impact on output performance. Unfortunately, the scale of this impact is unknowable, and the timing of the impact will be uncertain, but it *will* inevitably result in improvement in output performance. All the input and process improvements we make will be additive. Some, for example

Figure 8.2

a combination of improvement in procedures and training in those procedures can be synergistic, creating significant improvement.

What does all this tell us about setting improvement targets? One thing is certain – we cannot predict with accuracy or certainty what our future output performance will be, so setting output targets is something of an act of faith – it cannot *actually* be directly controlled. So what happens when we set such a target? The target becomes the focus, rather than the improvement. If necessary, people will hide accidents, distort data and apply penalties when things go wrong – all in the name of meeting the target. People's ingenuity knows no bounds when faced with such a challenge. Setting improvement targets for input and process factors is much more practical, because they can be accompanied by positive actions and measures which can be managed and controlled. People's energies can be directed at achieving things within their control, and their performance in achieving the goals can be monitored and feedback given. Forget about the outputs – they will look after themselves.

A further argument against output targets is the fear of not meeting them – and the effect of such a failure. You could argue that fear of not meeting the target is a motivator – which is no doubt true. But is it a *good* motivator? How much control do I have over the achievement of these imposed targets?

What if, part way through the measurement period, we go over the output target? Failure! What do we do for the rest of the measurement period? Panic? Give up? Whatever, the motivation is lost. We dug the hole and we fell in. What control did we ever have?

Targets and measures. Both are important, but more important is to know how to apply them. Use targets for input and process improvement. Use measures for inputs, processes and outputs, all in balance.

Output is subject to natural variation

What is not usually understood is that safety performance – generally as measured by injury rates – is naturally variable. If there is little significant change in your activities, and management style does not change, then safety performance will vary naturally between limits which are statistically determined by past performance. It might look something like Figure 8.3, where the accident rate is varying month to month about the mean rate, which is the average value of all the monthly rates.

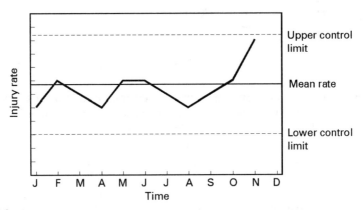

Figure 8.3

The upper and lower control limits are those which can be calculated using standard statistical techniques and, where the values fall between those control limits in a reasonably random way, their variation is quite natural. For this reason it is pointless to take specific action on a higher than average rate one month (e.g. November in Figure 8.3), even though it may vary considerably from the previous month's rate. If you look at the figures without the control limits as a guide, how do you know when things are moving out of control? In Figure 8.4, managers may well be forgiven for thinking that things are not only getting worse (pick your own trend from the data points!) and that November was the time to take some real, decisive action. Let's have a campaign! Then the figures go down again, following the natural cycle of variation, and managers breathe a little easier again. The campaign has had an effect! Well, it may have done at the time, but when the campaign is over, we are back to the same natural variation. Nothing has changed.

Take a look at Figure 8.5. There is clearly loss of control in October and November – worth investigating, since it looks as though it may be a special

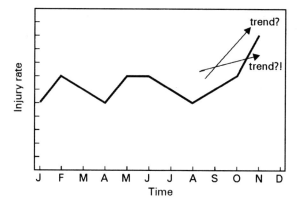

Figure 8.4

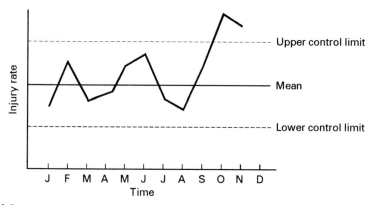

Figure 8.5

cause of variation. If you find that there is a specific cause – failures in training people in a changed operation, for example, then action to raise awareness and improve training to deal with this specific issue could be taken. Then, of course, you would want to establish the root causes of the failure to identify the need for training earlier.

This approach leads you from the *output* indicator (the accidents) to the underlying *process(es)* – in this case there was no process for retraining people when the operation changed – and maybe back to the broader *system* (for example the training strategy takes account of major changes in job role, but not for changes in operation within the same job role).

Exclusive focus on output measures encourages inaction or over-reaction

We have seen earlier the effect of lack of understanding of natural variation in output statistics, leading to complacency (when the figures are comparatively low) and panic (when the figures are comparatively high), when all the time all we were seeing was natural process variation.

In Figure 8.4 we could add to the injury rate axis 'level of panic' going up the scale and 'level of complacency' down the scale! For what can managers do with this data? It is enlightening how many theories can be generated from discussions about the ups and downs of accident charts with no control limits; for example, when accidents are down:

● are people working less hard?
● last month's initiative is working!
● maybe people are reporting less!
● maybe it's seasonal!

and, when accidents are up:

● we are not being tough enough on those having accidents – better start talking seriously with them!
● we don't know what it is, but *something's* wrong – better raise the profile!
● let's look for a cause (any old cause, so we can be seen to be taking action!)
● maybe it's seasonal!

The levels of frustration on the part of managers can be enormous, because they cannot gain control. Energy is dissipated by hopping from one issue to another – knee-jerk reactions to different situations. There is no focus on the things that really *will* make a long-term difference, because these things are not being measured and are therefore not getting attention. People become confused or complacent because they sense a lack of direction on the part of management.

How, then, do we set about measuring our performance effectively, which gives us a *balanced* view of progress and which allows us to have real control over that progress? Let's return to the point made earlier about outputs being the result of a combination of many inputs and processes. If we really want to understand how all of these factors are affecting our performance – and yes, we *do* need to understand them all! – we should have some measures which do that for us. With this in place we have a balanced measurement portfolio which gives us clear and useful information about our *inputs*, *processes* and *outputs*. And, although the relationship between them all is naturally complex, it gives us the opportunity for the first time to be able to correlate improvements in inputs and processes with improvements in output.

Now we'll take a closer look at options for measuring inputs, processes and outputs.

Output measures

We can start here with the measures you identified at the beginning of this chapter – the type of output measures that you yourself are already using. Let's look at those more closely. Figure 8.6 sets out the range of outputs normally identified with safety, health and environmental incidents, and all of which involve a cost to our business.

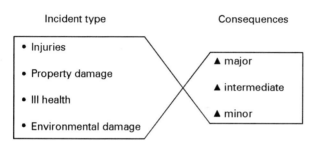

Incident type

- Injuries
- Property damage
- Ill health
- Environmental damage

Consequences

▲ major

▲ intermediate

▲ minor

Figure 8.6

For any of the incident types, there can be a range of consequences, which I have classed for simplicity as major, intermediate and minor. In terms of injuries, for example, a major injury would be death or serious injury, an intermediate injury would be one requiring professional medical treatment, e.g. stitches for a cut, and a minor injury would be a simple first aid case, e.g. a plaster on a cut.

We can measure the number or rate of incidents on a time scale for each of the incident types for each of their associated consequences. The injury-related chart, expressed in its most simple form, might look something like Figure 8.7.

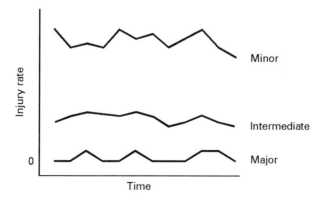

Injury rate

0

Minor

Intermediate

Major

Time

Figure 8.7

Your means of expressing the data for each of these three levels of injury may vary depending on their respective rates (number of injuries/1000 people/year, or number of injuries/100 000 exposure hours are typical rates). If the rate is very low (for example for major incidents) a simple data table or bar chart may suffice (Figure 8.8).

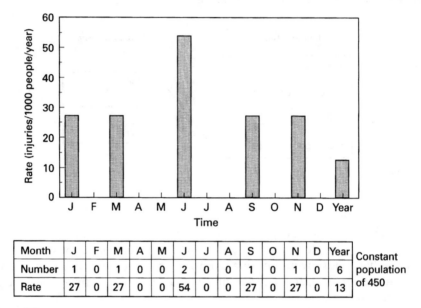

Month	J	F	M	A	M	J	J	A	S	O	N	D	Year	Constant
Number	1	0	1	0	0	2	0	0	1	0	1	0	6	population
Rate	27	0	27	0	0	54	0	0	27	0	27	0	13	of 450

Figure 8.8

Where the rate is greater – perhaps for the intermediate injuries, and almost certainly for the minor injuries – a control chart is much more appropriate. As we saw earlier, this approach helps us to see what the natural variation is, so that we can pick out the special causes of variation (e.g. October and November in Figure 8.5) and act on them. It also enables us to see where there is a statistically significant change in performance, for example as in Figure 8.9.

An alternative means of measuring trends in performance, especially where numbers or rates are low, is using cusums, which plot the accumulated sum of differences from target values. Cusums are particularly used to help detect small changes in the mean level of the variable being plotted, appearing as changes in the slope of the cusum graph. This is particularly helpful where the variable being plotted (e.g. injury rates) is low. In Figure 8.10, for example, it is very clear from the cusum chart that there is a change from July in year 1 and from September in year 2.

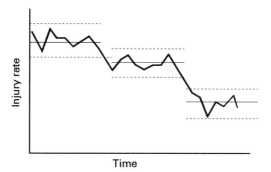

Figure 8.9

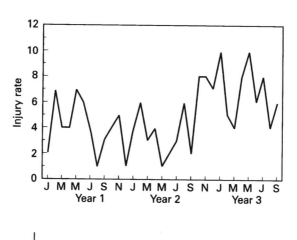

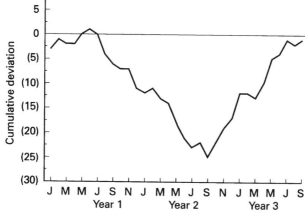

Figure 8.10

These are some measurement techniques for *actual* loss-producing incidents, but what about those incidents which do not result in loss? What have we learned already about accidents and their consequences, if not that consequences are often fortuitous? For example, the root causes of some intermediate incidents could never result in anything more serious. The root causes of some minor incidents or even no-damage incidents, could easily result in very serious incidents under slightly different circumstances – the rubber O-ring on the Space Shuttle *Challenger*, for example. How do we take account of this, since it is obviously important data?

One approach is to assess the *potential* of all incidents, *including* the no-damage incidents. Doing this really starts to tell us something about our potential for serious incidents, especially if we can get *all* the no-damage and minor incidents (which are easy to cover up or just avoid reporting) reported and analysed. You will know you are developing into a healthy cultural and safety management position when you are seeing the sort of trend in Figure 8.11, where the rate of reported injuries and the rate of reported no-damage incidents cross over. In this situation you are learning increasingly from no-damage incidents rather than from injury incidents, and are putting improvements in place *before* injury or damage occurs.

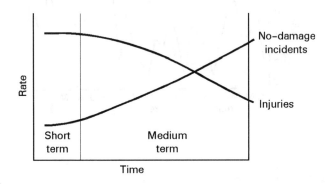

Figure 8.11

A longer term view of Figure 8.11 above is illustrated in Figure 8.12. Here your improvement efforts (including the encouragement of a culture in which people are encouraged to report incidents) start to show results in the short term – a start in the reduction of injuries and a rise in the number of no-damage injuries reported.

In the medium term, the trend continues with cross-over of rates. In the longer term, your continued efforts start to reduce the levels of no-damage incidents, reversing the upward trend, and the rate of no-damage incidents starts to decline also.

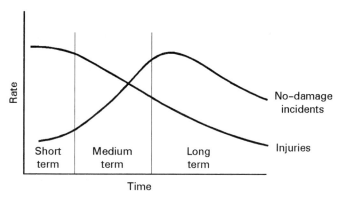

Figure 8.12

How can we assess the *potential* of incidents? Table 8.1 is an example of the way in which potential can be assessed, and the measures charted cumulatively through time (Figure 8.13). In Table 8.1, the *actual* severity of each incident in injury terms (both injury and non-injury incidents) is recorded, and also the *potential* severity, based on the worst credible event (taking into account all the circumstances of the incident, and making the assessment in a consistent way). As time progresses, the cumulative actuals and potentials may be charted, for example as in Figure 8.13, which demonstrates both the actual situation relating to injuries and the worst case potential situation.

How does this help? Well, this approach starts to tell us more about the incidents that are happening. It also helps us to focus on those incidents with greatest *potential*, rather than concentrate solely on the major and intermediate *outputs*. Having this focus helps us to get to the root causes of the more potentially significant incidents, helping us to probe more intelligently into the weaknesses in our *process* and *input* arrangements which are leaving us vulnerable to these incidents. Clearly, it will not tell us

Table 8.1

Reference	Actual severity	Potential severity
85/1/1	minor	intermediate
85/1/2	minor	minor
85/1/3	intermediate	major
85/1/4	intermediate	intermediate
85/1/5	minor	single fatality
85/1/6	no injury	intermediate

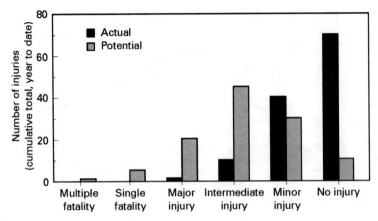

Figure 8.13

everything that is going wrong, but it will tell us much more, giving us a better opportunity to nip potential problems in the bud.

The other advantage is that when the *actual* level of injuries is very low, having a picture of the *potential* of incidents helps to prevent complacency!

Input and process measures

In Chapter 2 we examined assessing performance in terms of inputs and processes, and looked at a structured way of making this assessment based on an input–process–output model (Figure 2.3).

The approach used was based on a self-assessment, which concentrated on the following elements which relate directly to those in Figure 2.3.

policy	input	
strategy and plans	input	
standards	input	
training	process	
auditing	process	
organizational culture	input	management commitment
	process	consulting
		communicating
		reviewing progress
		recognizing
		leading and improving
resources	input	
results	output	

If, for each element, we have a check list such as those in Chapter 2, and add a 'scoring' system to each check list, a quantitative measure of progress may be maintained. Those elements which are accompanied by a detailed implementation plan may be monitored at a further level of detail by measuring progress to the plan. Examples are:

- translation of standards into local procedures
- training plans
- auditing plans.

We will look at these three examples in a little more detail.

Translation of standards into local procedures

Local operating procedures are based both on the requirements of company standards and any additional local standards which are specific to local operations. Preparing procedures, or updating them where they have become very outdated, is usually a large task and one which does not usually attract an army of volunteers! It is important, therefore, to treat the job as a project, with a timescale for completion, and measure progress to that timescale. If, because of the number of procedures involved, the timescale is relatively long, it is worth breaking down the stages of progression through the project, and measuring progress to those stages (see Figure 8.14) This has the advantage of being able to measure from an early stage in the project, not having to wait until later to start measuring the delivery of the finished product.

Training plans

Of all the safety training that is relevant to the work that people do, a small part will be critical to the safety of operations. Critical training is that small

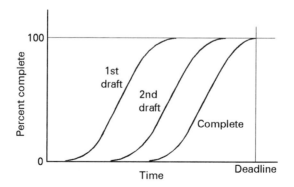

Figure 8.14

proportion of overall training which does two main things. The first is to deliver the fundamental principles of the organization on safety – setting the context for general behaviour and decision making which will steer individuals to err on the side of safety in new and untried situations. The second is to ensure high standards of competency in operating those procedures which, if not followed, could result in life-threatening situations – entry into vessels and work on live electrical equipment, for example.

Conformance to this type of training – both initial training and periodic retraining – should always be 100 per cent, and measurement should be made as easy as possible through readily accessible training records and registers of people who, following training (and often verification of the learning) have been authorized to perform critical tasks.

The remaining training – the greater part of it – may be delivered in may different ways; by formal 'classroom' sessions, by distance learning, by workplace team 'gen' sessions, and so on. As we saw in Chapter 6, the type of training should be consistent with learning needs, and the training and retraining needs of each individual planned. The delivery of this training and retraining should be recorded so that overall progress can be measured. This measurement may be made in different ways, depending on how close or far away you are from the target of 100 per cent accomplished training. If you are close, and have your training and retraining well under control, a simple bar chart would suffice. If you have a long way to go (e.g. Figure 8.15), you may need more detail. For example, you may wish to know how late is the training which has missed its scheduled date. If the pattern looks like Figure 8.15, you start to get a better picture of how far away you are from your targets, and you would seriously question whether resources or commitment are matching the demands of the training schedules – and whether the schedules themselves are reasonable.

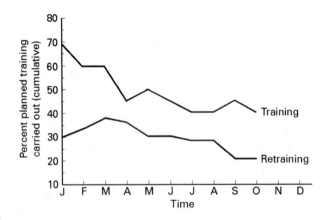

Figure 8.15

Auditing plans

Auditing is your vital organizational 'health check', so it is important to see how well auditing programmes are being implemented, and what conformance levels are being achieved. Both measures are needed: there is little comfort in all the audits being carried out to schedule if the conformance levels identified by the audits are continuously low!

A simple plot – as in Figure 8.16, for example – will help you track progress. The 'programme' line is the planned progression of audits, and the lower is the 'human nature' curve which will almost always be lagging behind. To save a panic as you approach the deadline, you could set your target curve to the dotted line!

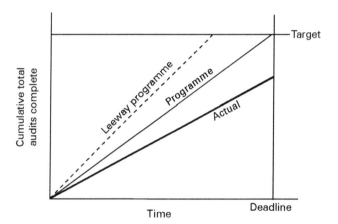

Figure 8.16

Figure 8.17 shows ways in which the conformance levels of audits may be illustrated. If there are a large number of procedures to audit, their auditing plan may be implemented over a long period – over a two-year cycle, for example – during which some of the more critical procedures will be audited several times and the least critical only once. Let us say that each audit receives a conformance rating of:

1 for complete non-conformance
2 for major non-conformance
3 for minor non-conformance
4 for total conformance.

Constructing a chart such as that shown in Figure 8.17 allows you to see a profile of the output of the audits over the audit cycle period (in this case two years). The bar chart illustrates the percentage of the total number of

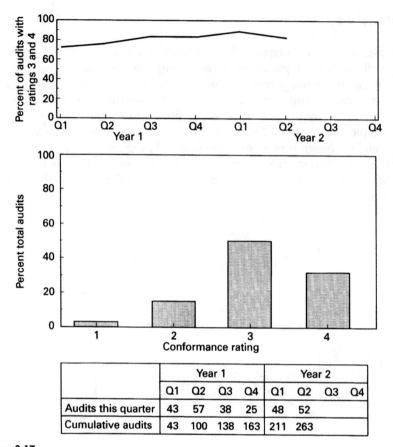

	Year 1				Year 2			
	Q1	Q2	Q3	Q4	Q1	Q2	Q3	Q4
Audits this quarter	43	57	38	25	48	52		
Cumulative audits	43	100	138	163	211	263		

Figure 8.17

audits in conformance ratings 1 to 4 up to the second quarter of the second year. The run chart (top) illustrates the percentage of audits that have a high level of conformance (either total conformance or minor non-conformance) quarter by quarter. This is an example of taking a broad overview of how auditing and its results are progressing, and on an audit cycle to audit cycle basis (in this case comparing one two-year cycle against the next), we would be looking for an increase in the percentage of total audits gaining conformance ratings 3 and 4.

You will gather from these timescales that we are looking at long-term measures. But remember that auditing *is* a long-term exercise and we are seeking progressive improvements in the long term. Resist, for example, comparing one month's conformance ratings against another; this will simply be comparing apples and pears, since different procedures will be audited from month to month, and there will inevitably be differences in levels of conformance from one procedure to another. Overall progress may only be accurately be gauged from one audit cycle to another.

Aside from the overview of progress just described, however, it is important to look for low conformance ratings in those procedures which are critical to safety of operations, and to identify trends in low conformance within individual operating units. Remember always to ask yourself what the data is trying to tell you. Often, problems in understanding data are more to do with the way we are interpreting and expressing the data than with the data itself – so feel free to experiment!

Then there are measures that can be made on any number of more specific issues which are important to overall progress. Examples are:

- attendance levels at scheduled safety training events
- levels of conformance to schedules in safety improvement plans
- time to complete incident reports.

You may well have already thought of more that are relevant to your own priorities. The aim here is to identify the internal standards that you want to achieve, and measure progress to those standards. There is nothing so designed to gain management attention as measurement to agreed standards. So, for example, if the aim is to achieve 100 per cent turnout to safety training events, measures of performance against this target, brought to the attention of senior management, will almost guarantee action.

The same applies for the turnround of accident reports – set a standard by agreement with management, then measure performance to that standard and bring to the attention of those managers who approved the standard. Performance will improve. These are important indicators because they relate to the 'softer' issues of organizational culture and management commitment.

The examples given are by no means comprehensive – they are shared from experience. You will be able to identify your own, but the principles behind the measures are similar, and usually very simple. If it is important, set a standard, measure against it and keep the profile high. What gets measured gets improved.

Feedback, publicizing and celebrating

We have discussed the advantages of a *balanced* measurement portfolio for tracking the performance of the organization as a whole. This approach should also be encouraged at the local level, where people can find measures of progress that are directly relevant to *them*. The broader organizational measures can be used as both encouragement and examples to local action. To help this move towards a measurement culture, it is necessary to give the organizational measures wide publicity, openly displaying progress to important organizational goals and celebrating when goals are reached. Publicizing your measures and celebrating success keeps the profile of your safety improvement programme high, reminding people of its importance to the organization and allowing people to see that real progress is being made – especially important where they are personally involved in safety improvement projects – and encourages local improvement initiatives. Local action and successes, too, should be publicized, helping to share improvement ideas and build success on success.

This overt measuring and celebrating success is an important part of organizational culture, and is seen in all companies which are achieving world class safety performance.The extent and style with which this is done varies between these companies, but it is there, and employees have no doubt about the importance the company attaches to achieving its safety, health and environmental targets.

Making measures work (pits, traps and tips)

There are lies, damn lies and statistics. It is very true that it is easy to distort data if you are intent on making it tell a particular story, but *we* are interested in managing our safety performance, so it isn't in our interests to distort the data which is trying to tell us something about that performance. Usually, we do not intentionally distort the data, but we do sometimes distort it unintentionally, and we do not always make best use of the data which we have. So, this section of the chapter gives a few examples and pointers to help to avoid those pits and traps.

Get down to actionable data

Remember why we measure in the first place. It isn't *just* to see how we are doing against the targets we set ourselves, but to identify where action needs to be taken to deal with the causes of unacceptable performance. For all the measures you take, consider what the measure will tell you about what action to take. It may, of course, tell you that no action is necessary – things are progressing satisfactorily. However, if the measure tells you that things are *not* satisfactory, but gives no indication of what action might be appropriate, this can lead to ineffective, poorly targeted action and management frustration.

The earlier example of the natural variability of injury rates is an example of ineffective management action. Here is another example from a chemical plant processing corrosive liquids. Figure 8.18 is the measure of the number of chemical burns on a month by month basis.

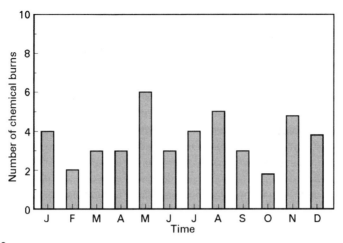

Figure 8.18

If I am the plant manager, what do I do? I know that the level of injuries is is unacceptable, in fact, because chemical burns can be serious, I would like to reduce the levels of injury right down towards zero. I certainly want to eliminate all serious chemical burns. What does Figure 8.19 tell me?

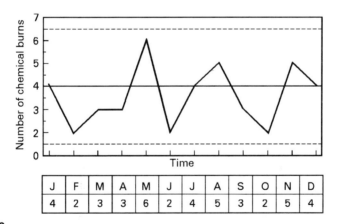

J	F	M	A	M	J	J	A	S	O	N	D
4	2	3	3	6	2	4	5	3	2	5	4

Figure 8.19

It tells me that there are no special causes of variation, and that to improve the situation I am going to have to work on the 'system', whatever that is. Where should I start? Well the accident investigation data tells me something about the underlying causes of these injuries, and if I chart them simply (see Figure 8.20), I start to understand more about what causative factors are playing a part in the accidents, and to what extent.

What does Figure 8.20 tell me? It tells me first of all that failure to wear personal protective equipment (PPE) is a factor in nearly all the accidents. It also tells me that in 50 per cent of the accidents, a failure to follow operating procedures results in exposure to the corrosive liquid, and that in 40 per cent of the cases this is due to lack of understanding of the procedures because of no training or poor training. It also tells me that in a small number of cases the use of contaminated PPE was a contributory cause.

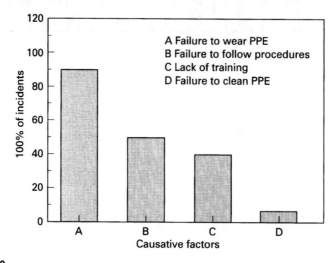

Figure 8.20

Now I am starting to home in on the issues on which I can take some action. I can identify the weaknesses in training arrangements for important operating procedures, strengthen them and introduce measures of compliance to training programmes (causative factor C, which should have an impact on causative factor B), and I can examine and improve arrangements for ensuring PPE is cleaned or discarded after use.

Caustive factor A is more difficult. What do I do about this? I dig deeper and what do I find about the factors that influence failure to wear PPE? They are:

● the PPE is uncomfortable and cumbersome
● the storage area for PPE is remote from where it is usually needed, and is often ill-stocked
● not wearing the PPE gets the job done quicker

- in the work scheduling, no time is allowed for preparing for the job and cleaning equipment afterwards
- when PPE is not worn, no-one remarks on the non-conformance
- it has become common practice *not* to wear PPE – there is some peer pressure not to do so.

Here are things I can now do something about, which will make a real improvement. I can directly influence the first four factors by arranging more convenient storage, more user-friendly PPE, and better work scheduling. I can also start to exert more positive leadership in the last two factors; neither I nor my supervisors will in future let any failure to wear PPE go unchecked.

Measurement by itself would not have got us down to *all* these important actions, but it can take us close enough for us to see what that last step needs to be. This use of measurement to analyse what is happening, and to identify positive and effective action to take, is really what measurement is all about.

Keep it simple

Measures and data can be mind-bogglingly complex, guaranteeing minimum understanding and maximum withdrawal. We might indeed have lots of data, and wonderful software programs to permutate the data in hundreds of different ways, but why drown people in it? We measure to understand and to be able to gain control. So why not do this by presenting simple pictures which tell us honestly what the important facts are.

Show the data with the chart

Anyone who has listened to politicians telling us that last month's figure on economic growth of 1 per cent is the start of an upturn as a result of the government's economic policies will have learned to have a healthy disrespect for interpreted data. Showing the source data with your graphical representation of it allows people to check the validity of your interpretation and gives them more confidence in it – and in you. An example of the added value of this approach is where source data is charted in a modified form – for example numbers of a particular injury type translated as 'percentage total injuries', or numbers of incidents expressed as incident rates. In cases like this, providing the source data with the graphical representation can tell people more about what the data is saying.

Look at Figure 8.21. What does it tell you? Time to panic!

Now look at the source data. From this it is clear that the increase is the result of a small (possibly natural) variation in injuries, but because it is a

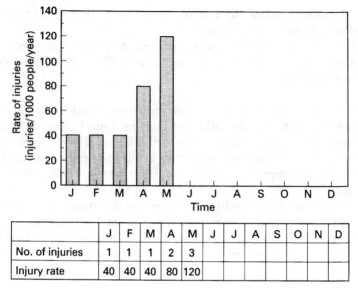

	J	F	M	A	M	J	J	A	S	O	N	D
No. of injuries	1	1	1	2	3							
Injury rate	40	40	40	80	120							

Figure 8.21

variance from a very low base figure, it makes the position – as illustrated in Figure 8.21 – look much worse than it actually is. The graph and its data tells us more about the need to use a different statistical technique than it does about our accident performance.

Other ways we can be fooled

There are many examples of data fooling us that things are getting better, getting worse, or stable, when the reality of the situation is quite different. We have to know on what basis the data is offered. For example, a transport company declares that its arrive-on-time record has jumped from 85 to 95 per cent. Good news for travellers! What it fails to say is that its criteria for 'arrive-on-time' has been changed from within 5 minutes of scheduled time to within 10 minutes of scheduled time. This is a shifty use of statistics, and tells us that we always have to question the underlying basis of the declared data – have the ground rules, or the standard, or the population on which the data is based been changed in any way?

We can distort our own data quite unintentionally. For example, measuring number of injuries in a population that is increasing or decreasing can give a distorted picture. Look at Figure 8.22. What does it tell us? Things are getting much better! But if we look at Figure 8.23 we see that we have a reducing population, and the *real* picture is shown on Figure 8.24 – essentially no change at all!

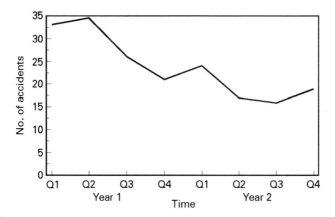

Figure 8.22

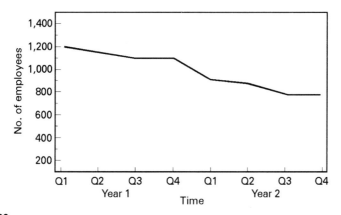

Figure 8.23

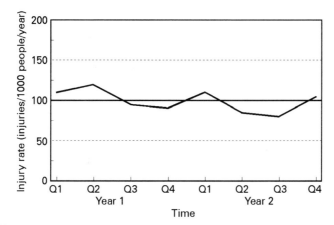

Figure 8.24

This is an example of a gradual change which can fool us. There are also examples of step changes that can be misleading, rather like the transport example. Where there is a reorganization, for example, and an operating section which has a particularly good or particularly poor accident performance moves from one accounting unit to another, the accident performance of those accounting units could be distorted – one could immediately look better, the other perhaps worse, when actually nothing has changed.

Because of the organizational structures we find ourselves in, we sometimes end up counting apples with pears in terms of work type – for example manufacturing with administration, when the difference in accident performance may be quite marked, the worst performer dominating the accident statistics. This has the effect of masking the *real* performance of the worst performer, often reducing the incentive to improve. And there is no incentive for the better performer to improve because its *true* performance is not being recognized.

There are more subtle influences on accident rates which are unquantifiable. We have to look for them by knowing our organizational shifts in behaviour. A common influence of this nature is the openness of reporting. As we improve the openness of our culture and encourage reporting, people report more minor injuries that previously have been hidden, and this can have a distorting effect on statistics. This can neutralize a downward trend in accidents which is the outcome of the safety management improvements that are being made. However, this will only happen for so long, until the level of reporting settles out consistent with your new culture. Moreover, the trends in the more serious accidents (which are more difficult to hide) will give you an output indication of improvement which is less affected by improvements in cultural openness.

Train people in simple measurement techniques

We may think that because of our background education, where measuring things features strongly, and in some cases because of the nature of our occupation, that measurement of performance in the things that we do is second nature. Actually it is not. In one organization, measurement became a new 'in thing' and everyone was encouraged (in some cases forced) to have at least one measure of progress in their work. The result? In the great majority of cases people measured those things that were least threatening, and often least relevant to the core of their work, so that the measures were ineffective in helping to achieve improvement. A lot of energy was wasted because people didn't have the understanding and basic skills to measure effectively.

The message here is just the same as those messages in Chapter 2. We can gain tremendous leverage by empowering people to make improvements themselves, and enabling them to measure effectively is just as important a tool for them as it is for those few who are measuring overall organizational progress. Training people in simple measurement and problem solving tools (often needed to take an analytical approach to the outcome of measurement) is not difficult or very time consuming, and can have significant payback.

No measure has to be forever

I said before that measures help us to assess our performance and its progress. They also help to drive action – to move us towards our vision of what we want to achieve; what sort of organization we want to be. Eventually, if we are serious about achieving that goal, we will get there, and some of the measures we have used to help us along the way will no longer be necessary. We will always need some *maintenance* measures to let us know that we are not slipping on those absolutely essential 'process' factors – training and auditing, for example. But some measures will be redundant because the new way of working, which those measures helped us achieve, is now a way of life.

In this situation measures are still needed to identify and get down to the root causes of problems, and to go forward to the next stage of improvement. For as I said before, we are never dealing with a static situation. Organizations, and the external influences on them, are dynamic, so it will always be necessary to seek to improve – even to stand still! Complacency always results in decline – in safety performance as in any other aspect of business performance.

You should not have been surprised to notice that on the whole the pitfalls and bear traps of measurement are associated with *output* measures. Again, this reminds us how important it is to have a *balanced* measurement portfolio. Measures are powerful. They drive management action. So we should use them intelligently and responsibly, helping us to direct our energies most effectively in gaining successful improvement.

Conclusions

Let's face it. Being in business is not simple. But since we *are* in business, it's in our interests to stay afloat, and minimizing losses – money off the bottom line – is worth the effort involved. Remember that all we have covered in this book is not established overnight – it is a steady and deliberate process of improvement which will not just improve your safety performance. It will, if managed well, improve *all* your operations. As I said at the beginning, good safety management is good *anything* management.

Further reading

Here are some suggestions for further reading. It is a relatively short list for three main reasons. Firstly, there are simply too many relevant references to include them all. Secondly, if I listed them all, you would not be tempted to read *any* of them. Thirdly, of all the texts I have read which are relevant to the context of this book, these are the ones which I have personally found helpful. That isn't to say, of course, that there aren't other excellent texts: no doubt there are – it's just that I haven't had the opportunity to read them.

Chapter 1

Kletz, T (1989) *Learning from Accidents in Industry,* Butterworth-Heinemann. (Good on root causes, common sense, human behaviour.)
Krause, TR (1990) *The Behaviour-based Safety Process,* Van Nostrand Reinhold. (Interesting insights into human behaviour and how to influence it by focusing on the consequences of behaviours.)
Report of the Presidential Commission on the Space Shuttle Challenger Accident (1986).
Herald of Free Enterprise, Dept of Transport (1988), HMSO.
Investigation into the Clapham Junction Railway Accident (1989), HMSO.
The Public Enquiry into the Piper Alpha Disaster (1990), HMSO.
Successful Health and Safety Management, HSE (1991), HMSO. (Crisp messages about the essentials of health and safety management.)

Chapter 2

Byham, WC (1988) *Zapp! The Lightning of Empowerment,* Century Business. (Entertaining and instructive messages about empowerment.)

Camp, RC (1989) *Benchmarking: The Search for Industry Best Practices that Lead to Superior Performance,* QASQ Quality Press. (Recognized leader-in-the-field text on benchmarking practice.)

Developing a Safety Culture (1990) CBI. (Sound, practical pointers on personal behaviours and organizational culture based on industry experience.)

Deming, WE (1982) *Out Of The Crisis,* Cambridge University Press. (Classic text from the grand master of improvement in business practice.)

Kohn, A (1993) *Punished By Rewards,* Houghton Mifflin Company. (Sobering and instructive text on recognition and rewards.)

Scholtes, PR (1988) *The Team Handbook,* Joiner Associates. (How to motivate teams and make projects work.)

Chapter 4

Guidelines for Chemical Process Quantitative Risk Analysis (1989) American Institute of Chemical Engineers. (A professional's text, but useful stuff for the non-specialist if you are able to pick and choose.)

Kletz, T (1991) *An Engineer's View of Human Error,* Institute of Chemical Engineers. (Simply stated and readable text on human error.)

Chapter 6

Bambrough, J (1993) *Training Your Staff,* The Industrial Society. (Short and simply presented overview of planning, giving and evaluating training.)

Honey, P and Mumford, A (1983) *Using Your Learning Styles,* Peter Honey. (Concise account of learning styles and how to improve them.)

Chapter 7

International Safety Rating System (1978) International Loss Control Institute, Institute Publishing.

Chapter 8

Wheeler, DJ (1993) *Understanding Variation: The Key to Managing Chaos,* SPC Press Inc. (Classic, simply presented text on variation and process control: an essential read for any manager who is serious about performance measurement.)

Index